ISBN 978-94-015-1685-3   ISBN 978-94-015-2829-0 (eBook)
DOI 10.1007/978-94-015-2829-0

Reprinted from Bibliographia Genetica XVIII

Bibliographia Genetica XVIII (1959) 101–166.

MALE STERILITY IN FLOWERING PLANTS

by

S. K. JAIN

Dept. of Genetics, Univ. of California,
Davis, Calif., U.S.A.

(*Received for publication July 24, 1958*).

CONTENTS

I. INTRODUCTION

The phenomenon of male sterility received ample recognition at the hands of many early students of sex in plant kingdom and eversince both taxonomists and evolutionists have shown their interest in such associated problems as those dealing with the mode of reproduction and breeding structure of a plant group or the diversity of floral patterns and their teratism. GÄRTNER (1844) and DARWIN (1890) had frequently encountered the occurrence of contabescent anthers in several members of the families *Caryophyllaceae, Ericaceae* and *Liliaceae*. This anomalous character was attributed a definite role in the evolution of dioecism which is an obligate crossfertilizing adaptation in amphimictic species. Since then a vast amount of literature recording cases of sex reversal, sex suppression and other types of male sterility has accumulated.

The male sterility phenomenon however assumed its real genetic significance after the pioneering researches of BATESON and his associates (1908 and earlier) on a recessive monofactorial case in sweetpeas, of SALAMAN (1910, 1912) on dominant male-sterile character of potatoes (several species of *Solanum*) and of CORRENS (1908) on several gynodioecious species where the cytoplasm appears to govern its inheritance. In fact male sterility has been one of the few wellestablished examples of cytoplasmic inheritance. Of still greater interest are the various modes of its expression that are, in general, characterized by the irregularities of microsporogenesis and of pollen mitoses. A cytological study of different male-sterile cases occurring in nature has revealed a precise stepwise inhibition of those normal processes involved in the formation of male sex cells. This may be shown by future research to simulate a biochemical mechanism found by modern geneticists in lower organisms.

Plant breeders and horticulturists, on the other hand, first realized the importance of male sterility in connection with the lowered seed or fruit production as a consequence of the lack of optimum pollination conditions (CLARK and FRYER, 1920; ARMSTRONG and WHITE, 1935; VALLEAU, 1918; KVAALE, 1927; and many others). In order to minimize the influence of poor pollen producers, systematic planning of a field planting or an orchard might be necessary. Above all, the important reason for its wide popularity is its potential use in the commercial production of hybrid seed. A male-sterile plant is an effective female for a crossing program and its employment renders the laborious procedure of emasculation superfluous. JONES and EMSWELLER (1937) and STEPHENS (1937) were probably the first to outline a scheme of its application in onion and sorghum respectively. However certain specific problems have limited the practicability and hence some of them will be mentioned in a later section (Section VIII).

Lately several attempts have been made to obtain some chemical method for the artificial induction of male sterility. However, for a successful approach to the induction problem a thorough understanding of various pathways that bring about pollen abortion in naturally occurring cases, would be of great value. It may be hoped that data from these induction experiments alongwith the available biochemical analyses would lead to the emergence and elucidation of useful information of both theoretical and practical interest.

The purpose of this review is primarily to attempt a survey of the phenomenon of male sterility as occurring in nature and in experimental studies with a view to emphasize upon a coherent picture of all its phases together. The general notion that it is merely a byproduct of the eversportive nature of biological world, should be replaced by an adequate estimate of its potential value as a subject of further research. It might be well pointed out that the previous reviews as those of CORRENS (1928), LOEHWING (1938), ALLEN (1940) and EDWARDSON (1956) dealt with more or less individual aspects rather than from a generalized viewpoint.

II. DEFINITION AND SOME GENERAL FEATURES

Before proceeding any further, it would be appropriate to make up a formal definition of male sterility. Ordinarily speaking, sterility (or selfsterility) denotes a condition in which unless specified otherwise, both pollen grains and ovules are nonfunctional and usually a certain amount of morphological and cytohistological deterioration may accompany it. The instances of selfincompatibility, in contrast, have functional parts of the flower and gametes normally developed. Male sterility is a more specific term and as defined by DORSEY (1914), represents "the condition resulting from defects leading upto the nonformation of pollen, or lack of functional power in it when formed." This important point in considering both quantity and quality of pollen with reference to sterility was explicitly stated by BEADLE (1932 b) and CHILDERS (1952). Accordingly a male-sterile individual has either no pollen formed, or pollen inviable and incapable of effecting fertilization under normal conditions. It is an obvious implication that the ovular fertility is completely or almost completely unimpaired in a male-sterile individual.

At this point a distinction need be introduced so as we may identify male sterility and pollen sterility separately. The former is inclusive of the latter and in its widest sense, it denotes the condition where an individual has only female parts functional. This meaning alone can justify the classical interpretation of the females among gyno-dioecious species as malesteriles. In a similar manner, the inclusion of the instances of disbalanced sex ratios in a population and of sex reversal or sex suppression insofar these changes are unilateral and result in deficiency of 'maleness', seems feasible. Therefore a formal definition of male sterility shall denote by it all cases involving (i) deficiency of male individuals in a dioecious strain, (ii) absence or atrophy of male organs in a normally bisexual plant, (iii) failure to produce normal sporogenous tissues in stamens, (iv) inhibition at

various stages of pollen development yielding incomplete or imper-
fect pollen, or finally, (v) failure to mature, dehisce or to function,
when placed on a compatible stigma. Pollen abortion and pollen
failure (iv and v above) may be considered together as pollen sterility.
Thus it follows that the operation of pollen sterility will be readily
recognized only where the flowers are perfect and homomorphic, the
plant is normal in other respects, compatibility factors are absent
and self- or crosspollinations can be easily made.

In most genetically controlled instances of male sterility, the
stage at which the pollen development breaks down is highly specific
for a particular case and usually all processes upto that stage are
apparently normal. On the other hand, cases involving environmental
factors as the chief cause of sterile condition, exhibit considerable
amount of variability in these respects. The pollen sterile character
may vary in expression from plant to plant or even within a plant
between the different flowers or between the anthers of same flower.
That howfar the internal primary controls and thereupon super-
imposed secondary metabolic control mechanisms, or the timing and
rate differences of gene action contribute to this characteristic variability, is very little known at present. In most instances, however,
relatively simple inheritance that is monogenic or digenic, have been
met with than would be expected for a meristic character.

The characteristic variability of this phenomenon however calls
for some arbitrary scheme of classifying several different grades,
viz. high, medium and low. KOOPMANS (1951, 1952) and OEHLKERS
(1952) described two to as many as six classes of anther reduction,
whereas FUKASAWA (1956) classed various types of pollen degeneration
into five grades depending upon the stage of its onset and the phenotypic appearance of pollen. There are differences in opinion, for instance, about the limits of complete pollen sterility grade. KLEY
(1954) considered it to be complete when only 0.1 to 0,5 percent
pollen is functional. However, for all practical purposes even upto
5% fertility may not be considered as any serious deviation from
'complete' grade.

III. MODES OF ORIGIN

Male-sterile individuals have arisen in nature under a variety of circumstances that might perhaps be considered as some of its regular sources. Of much greater importance to us are, however, the cases involving our familiar genetic materials. FRANKEL (1940) reported a rather high frequency of male-steriles occurring among the artificially inbred populations of several species of *Hebe* and in fact, he observed that smaller sized populations tend to favour their occurrence as would be expected for there being greater chances of sibmating or some other kind of inbreeding that helps reveal many recessive genotypes in homozygous condition. Furthermore, according to FRANKEL, male sterility once established, would depend for its maintenance upon such factors as "(a) balance between cross- and selffertility factors; (b) relative advantage of the two breeding systems; (c) population density; and (d) flowering mechanism of the species in question". In wellbuffered genetic materials, an equilibrium may have been set up so that factors causing instability of meiotic behaviour (DARLINGTON, 1937) or of sex expression (SHIFRISS, 1956) would be absent.

A number of examples may be given to illustrate the rise of male-steriles among artificially inbred materials (*Secale*, DAVIDSON et al, 1924; *Morus*, SCHAFFNER, 1929; *Zea*, CLARK, 1942). MÜNTZING (1932) expressed the opinion that perhaps all typical allogamous species are characteristically low in pollen fertility. STOUT and CLARK (1924) concluded that the whole race of cultivated potatoes is deficient in 'maleness'. On the other hand, a systematic and extensive search may demonstrate a regular occurrence of male-sterile plants among the natural or artificial populations of a self-fertilized species as well (LEWIS, 1941; ALLARD, 1953, in limabeans). Theoretical considerations of its possible evolutionary importance are given elsewhere (Section IX).

Male sterility of one form or another has independently occurred among several apomictic species. An extreme case with marked anther degeneration was reported in *Hieracium excellens* (OSTEN-FELD, 1906) in which diploid parthenogenesis is known to be the prevalent mode of reproduction. In *Paspalum notatum*, BURTON (1948) could make ingenious use of a male-sterile plant for facilitating the study of its apomictic method of reproduction. Since this species is essentially selfpollinated, the male-sterile alone provided an opportunity for natural crosspollination. Most of this seed from male-sterile clone gave exactly maternal plants and cytological studies further confirmed the unreduced pseudogamous reproduction. Male-sterility and white stigma colour, both recessives, were found useful as marker genes. Instances of male sterility have also been reported in *Potentilla* (ROSCOE, 1927b), *Alnus rugosa* (WOODWORTH, 1930) and in *Parthenium argentatum* (POWERS and GARDNER, 1945). In *Saccharum officinarum*, JANAKIAMMAL (1941) reported a male-sterile having arisen in the progeny when pollinations were made with the pollen of *Imperata cylindrica*. It was shown to be a derivative of diploid parthenogenesis.

HAYES et al (1955) observed that a number of plants with all maternal characteristics that have occurred in the progeny tests of crosspollinated species, may have resulted from apomictic reproduction. STRASBURGER (1905) had originally put forth a hypothesis that male sterility may be a direct cause of the origin of apomictic behaviour in some species. On contrary, MÜNTZING (1932) obtained evidence from his studies in *Poa* that this coextensive occurrence of these two phenomena was merely fortuitous. It may be pointed out that many of the apomictic species that have been cytologically investigated, exhibit aberrant types of microsporogenesis substituting the normal meiotic divisions.

A third mode of origin of the male-sterile condition can be attributed to hybrid ancestry, that is, male sterility may arise following hybridization between or even within species. The most likely explanation for these cases assumes some kind of disharmony between the nuclear and the cytoplasmic components of parents (See section Vc). For instance, CLAUSEN (1926) in *Viola tricolor* × *arvensis*, SELIM (1931) in *Oryza indica* × *japonica* hybrids, SAUNDERS and STEBBINS (1936) in several interspecific hybrids of *Paeonia*, JANA-

KIAMMAL (1942) in *Saccharum officinarum* × *Narenga porphyrocoma*, BURTON (1943) in *Paspalum urvillei* × *malacophyllum*, SCHNELL (1948) in *Solanum demissum* × *tuberosum*, KOOPMANS (1951, 1952–1959) in *Solanum rybinii* × *chacoense* and its reciprocal and BROCK (1954) in *Lilium* hybrids recorded varying degrees of anther suppression or pollen abortion. In many species of *Wisteria*, ROSCOE (1927a) observed frequently pollen aborting at or soonafter the tetrad stage and he attributes this to the heterozygosity present in this material. As an indirect cause of the irregularities of pollen development, the numerical or the structural hybridity with reference to the chromosome complement have been found in a number of cases (LONGLEY, 1926, in *Fragaria;* BLAKESLEE and CARTLEDGE, 1927, in *Datura;* HEILBORN, 1932, in *Malus;* BHADURI, 1942, in *Rhoeo*). In the case of *Ranunculus acris*, SOROKIN (1927) showed the predominance of certain female forms to be related to various unbalanced types of changes in its chromosome complement.

STEPHENS (1942) found that some colchicine-induced tetraploids of *Gossypium arboreum neglectum* had 40–50% ovular sterility alongwith almost complete pollen sterility. BEASLEY (1942) obtained similar results with the allotetraploid, *Gossypium arboreum* × *thurberi*. It appears from these papers that in this genus, induced polyploidy tends to be accompanied by high amount of pollen sterility.

EAST (1932) carried out brilliant researches on the self-sterility alleles in many genera including *Nicotiana*. He was able to show an important role these genes play in the normal pollen production. *N. Sanderae* has two S alleles which when present in the cytoplasm of *N. Langsdorffii* give pollen sterility. Another example of an association between these two breeding systems was provided by JENKIN (1930) in *Lolium perenne* which LEWIS (1942) has interpreted as having involved a plasmatic factor as well.

Among many others, the external environment is an overall important factor in normal pollen development or even the general sex expression in many species. Thus the occurrence of male sterility may be found associated with cyclic or seasonal fluctuations in the sex ratios, reproductive activity or the gametic survival. Some of these causes of variation could be under internal biogenetic control (e.g. in *Cleome speciosa*, STOUT, 1923; in cucurbits, SHIFRISS, 1945). It is in this context that the question of explaining a relatively much rarer occurrence of female steriles may be raised. According to DAR-LINGTON (1946), "the differential tempo of development in haplophase gives differential emphasis on one sex (generally female) so that monoecy is led to femaleness alone."

LOEHWING (1938) has discussed at considerable length the physiological aspects of the sex problem. He considered the hypersensitivity of pollen or its developmental stages to the environment to be attributable to (a) the usual peripheral position of the male organs; (b) their requirements of longer photoperiods; (c) the higher oxidation potential associated with higher contents of sugars, phosphatase and oxidase in the staminate zones; (d) greater localization of various nutrient components. Experimental evidence has yet to be obtained in support of these assumptions.

IV. TYPES OF MALE STERILITY

In view of the definition of male sterility given earlier, a tentative scheme of classifying different types of naturally occurring male-steriles may be based primarily on two criteria, namely the phenotypic identity and the cause of incidence. Accordingly, following main types may be recognized:

1. *Teratological*. This class will include instances of both sex suppression and sex reversal from male into female (pistillody = transition of stamens into pistils). Among the progeny of certain dioecious strains, there may arise alterations in sex ratios towards the majority of females. The underlying causes might be several such as the differential survival of females or a developmental reversion in sex organs. The socalled gynodioecious species comprise females and hermaphrodites in a definite proportion. CORRENS (1908, 1926, 1928) carried out pioneering investigations in many of them. In *Cirsium oleraceum, Satureia hortensis* and *Silene armeria*, he found that both females and hermaphrodites bred true, and since repeated backcrosses to replace the entire genom of a parent by another did not change the true breeding behaviour of females, his findings lent support to the hypothesis postulated by WETTSTEIN (1924) that some factor of cytoplasmic origin suppressed the anther development among the progeny. In other gynodioecious species having both female and hermaphrodite forms segregated in progeny, no genetical explanation could be found. However it became quite evident that although the sexes may show sharp separation from each other, some elaborate genetic mechanism is often involved.

Cases of only partial reversal or suppression of sex elements exhibit many sorts of intersexes that often present the problem of proper classification and nomenculature. At least thirty different intergrading forms were recognized in *Plantago lanceolata* (CORRENS, 1908) and about fifteen of them in papaya (STOREY, 1953). We shall not attempt to discuss them here mainly because very little is known about them from the genetic viewpoint. However, mention must be made of castor, a typical monoecious species, where pistillate flowers are borne on the upper branches of the raceme and staminate flowers on the lower branches. The ratio between pistillate and staminate flowers in a raceme serves as an index of the sex behaviour of a plant. SHIFRISS (1956) recorded several cases of sex reversals among many natural populations and inbred races which he attributes to a general instability as well as to a genetic control.

Numerous instances of pistillody are on record in literature. The heritable cases may appear through a gene mutation, or following a cross through various genic, chromosomal or physiological disturbances. In a peculiar form of *Cheiranthus cheiri*, described as 'gynanthus' by DECANDOLLE, the androecium is replaced by a cylindrical column bearing supernumerary pistils. CHITTENDEN (1914) reported a 3 : 1 ratio in the F_2 of normal $\times$ gynantherous cross whereas SIRKS (1924) obtained a 358 normal : 50 gynantherous ratio explainable on the basis of a three gene assumption. In *Silene*, CORRENS (1928) found a monofactorial recessive form designated as 'polycarpellata' in which both petals and anthers are transformed into carpels. Several examples of pistillody operated by Mendelian factors are available in *Zea mays* (WEATHERWAX, 1925), *Ipomoea* (CORRENS, 1926), *Primula officinalis* (DAHLGREN, 1932) and in *Papaya* (STOREY, 1953). In *Prunus*, SMITH (1927) reported pistilloidic forms arising from interspecific crosses. Among nonheritable cases may be mentioned those found in *Hypericum nudiflorum* (RHEDER, 1911), in *Morus* and *Cannabis* (SCHAFFNER, 1927, 1929), in wheat and wheat-rye hybrids (LEIGHTY and SANDO, 1924; BLEDSOE, 1929) and in barley (GREGORY and PURVIS, 1947). As mentioned already, a common feature of all of them is the development of supernumerary carpellike structures, sometimes even functional, in place of part or whole of the androecium. STOW (1930) and MOL (1933) reported in hyacinths a different form of sex reversal where certain physiological stimuli seem to cause the duplication of generative nuclei in pollen so that they are transformed into embryosac-like giant structures. On basis of similar observations, SCHÜRHOFF (1922) had postulated that a male potency factor is preponderant in the vegetative point of the anthers.

The suppression of male sex becomes extreme in the form of anther-less mutants as those reported in sorghum (KARPER and STEPHENS, 1936), maize (HADJINOV, cited by EMERSON et al, 1935), tomato (RICK, 1945) and in tobacco (RAEBER and BOLTON, 1955). Or otherwise, anthers may reduce to staminodia (LONGLEY, 1926, in *Fragaria*), straplike appendages (SAUNDERS and STEBBINS, 1936, in *Paeonia* hybrids) or vestigial structures (OEHLKERS, 1938, in *Streptocarpus;* RAMAN, 1955, in *Jasminum*). The occurrence of disproportionate sex ratios has been severally reported (for instance, SCOTT, 1933, in bushpumpkins; WHITAKER, 1931, in many cucurbits).

2). *Anatomical.* CRANE (1915) in tomato and DORSEY (1914), SUSA (1927) and OLMO (1943) in grape found individuals with reflexed stamens while the stigma extruded normally. This anatomical defect results in nondehiscence of pollen although formed abundantly. ROEVER (1948) described an interesting mutant in tomato. wellknown as 'John Baer' male-sterile, in which the anthers do not dehisce due to the absence of an anther pore. Similarly, in a functional male-sterile of brinjal, JASMIN (1954) found that anthers developed normally except for the pore at the anther tip. In *Trifolium repens*, ATWOOD (1944) reported a case with anthers too hard to burst whereas in the 'vestigial glume' mutant of maize, first discovered by SPRAGUE (1939), the outer glumes being atrophied, anthers wither off prematurely and by the time of anthesis, no functional pollen grains are left. In some irregular situations, the entire whorl of stamens becomes a hard central column so that again there results nondehiscent condition. (RAMAER, 1935, in *Hevea*).

3). *Developmental.* Under this category, many physiological as well as genotypically controlled cases often expressed as pollen abortion or failure, may be included. Of course it is difficult sometimes to distinguish them from each other experimentally. Investigations by TAPLEY (1923) on the fruiting habit in squash, by MILLER (1929) on factors determining the seedstalk development in cabbage, by CURRENCE (1933) on the patterns of nodal sequence in cucumber and by KÜHN (1937) on physiology of pollen development in *Matthiola* have

established a number of forms of physiological pollen sterility. NA-
KAMURA (1943) expressed the opinion that during the period characterized by the appearance of large vacuoles pollen grains suffer from
a temporary nutritional deficiency for certain metabolic reasons.
Any slightest disturbance in the physiology of an individual may
therefore exert considerable influence upon the pollen development
during this critical period. In many *Oenothera* species, RENNER (1919)
considered periodical fluctuations in pollen fertility to be due to changes in starch reserves under genetic control.

Additional papers that reported male-sterile condition to be brought
about by some inherent physiological disturbances include those of
KOSTOFF (1930) on *Nicotiana*, JANAKIAMMAL (1942) on *Saccharum*,
ARTSCHWAGER (1947) on *Beta* and of CRANE and THOMAS (1949) on
Rubus. It may be inferred that nutritional factors seem to be many
and also common.

4). *Environmental*. In this supplementary type, the emphasis is
shifted from internal environment of a plant to its external conditions
with reference to the primary cause of the origin of male-sterile character. A number of potato workers have showed the predominant role
of atmospheric temperature in normal pollen development. (YOUNG,
1923; FUKUDA, 1927; STOW, 1927). MOL (1933) confirmed this observation in hyacinths that bloom early in the season.

From their studies in strawberry, VALLEAU (1918) and DARROW
(1927) concluded that light intensity and soil conditions largely
determine the sex expression in this genus and when floral parts are
affected, stamens are first to abort, next petals or sepals whereas
pistils are last. The results of NÖRDENSKIOLD (1945) in *Phleum pratense*, of NAKAMURA (1936) in *Impatiens Balsamina* and of BURTON
(1943) in *Paspalum notatum* merit special mention in this connection.
Often routine pollen counts reveal the presence of daily variations in
microsporogenesis and fertility which is a manifest result of environmental variations. (viz., LONGLEY and CLARK, 1930, in potato;
MICHAELIS, 1954 and earlier, in *Epilobium*). Some of the reports on
this type of male sterility appear incomplete or inconclusive as yet
although in future a genetic mechanism may possibly be found. For
instance, MARTIN and CRAWFORD (1951) mentioned two types of
pollen sterility in *Capsicum frutescens*. Of these one appearing in

field conditions is stable whereas the other appears only under green-house conditions and recovers when plants are taken to field.

5). *Unclassified*. Certain types of male sterility occurring in the genus *Hebe* (FRANKEL, 1940) and among *Musa* hybrids (DODDS and SIMMONDS, 1946) did not allow any evidence for either its mode of origin or inheritance so that apparently these instances await further analyses before an adequate evaluation is found possible.

V. INHERITANCE

Based on genetic data, all heritable cases of male sterility may be categorized as either genic, cytoplasmic or gene-cytoplasmic. Table I summarizes their known examples as occurring within a species whereas cases metwith among the interspecific hybrids are listed in Table 2. It is customary to denote the presence or absence of a plasm factor by + and — signs respectively. In columns of nuclear inheritance (Table I), the number of dominant and recessive genes controlling the male-sterile character are indicated. It will be noted that in some cases, there was reported an indication of genetic control, although exact information as to the number or kind of factor is missing. Since the genetic information in most instances appears either preliminary or identical, tabulation is considered all that may be necessary for our purpose here.

(A) *Genic male sterility.*

There are abundant examples of male sterility resulting solely from gene action. An overwhelming majority is known to involve a single recessive factor. A case of dominant gene governing pollen sterility was first reported by SALAMAN (1910) in several species of *Solanum*. Contrary to his earlier conclusion of a single factor pair, SALAMAN and LESLEY (1922) found it to be inherited on a complex pattern. Two other cases of dominant Mendelian control were reported in *Centhranthus* (SOBRINKO, cited after KLEY, 1954) and in *Coleus* (RIFE 1948; FORD, 1950). Thus it appears that mutability of the loci concerned with male sterile character falls in line with many others showing tendency of recessive degeneration. In fact, the recessive mutants would also have far better chance of survival in natural populations for a few generations after their first occurrence since the expression of lethality is sporophytic in most cases, that is, the heterozygotes are normal male-fertile. Exceptions to this general rule

TABLE I. Genetic data on known cases of heritable male sterility.
(Elaborated from KLEY, 1954).

S. No.	Material	Author	Plasm	Gene dom.	Gene rec.	Remarks
1.	*Allium cepa*	MONOSMITH	+	0	1	Pollen abortive. Complete sterility.
		JONES et al	+	0	1	
		FOSKETT	+	0	1	Originated in intraspp. crosses.
2.	*Alopecurus mysuroides*	JOHNSSON				Indication of genotypic control.
3.	*Anthirrhinum majus*	BAUR	—	0	1	Four nonallelic mutants.
4.	*Aquilegia vulgaris*	KAPPERT				Involve anther suppression Androecium reduced.
5.	*Beta vulgaris*	HOGABOAM	+	0	1	Anthers shrivel.
		OWEN	+	0	2	
	,,		—	0	1	Two factors, a_2 and a_3.
	,,		+	0	0	In crosspollinated material.
6.	*Capsella bursapastoris*	SHULL	—	0	1	pollen-lethal, sp. Close linkage with "subrhomboidea" locus.
7.	*Centhranthus*	SOBRINKO	—	1	1	(Cited after KLEY).
8.	*Cheiranthus cheiri*	CHITTENDEN	—	0	1	Gynantherous mutant.
		SIRKS	—	0	2(3)	
9.	*Cirsium oleraceum*	CORRENS				Gynodioecious.
10.	*Citrus*	NAKAMURA	—			Genotypic control.
11.	*Coleus*	RIFE	—	1	0	Linkage with leaflobe gene
		FORD	—	1	0	
12.	*Cucumis melo*	BOHN	—	0	1	(cited after KLEY)
13.	*Cucurbita pepo*	SHIFRISS	—	0	1	Androecium degenerated in bud stage.
14.	*Cucurbita maxima*	SCOTT et al	—	0	1	
15.	*Dactylis glomerata*	MYERS	+	?	0	Microspores abort. Modifying factors present.
16.	*Daucus carota*	WELCH et al				Genic but exact number not known.
17.	*Echinocloa*	AYYANGAR et al	—	0	1	Stamens do not exert.
18.	*Eleusine coracana*	,,	—	0	1	Pollen grains agglutinate. No dehiscence.
19.	*Godetia Whitneyi (Clarkia)*	HIORTH	+	0	1	Involves sex reversal.
20.	*Hebe townsoni*	FRANKEL	—	0	1	
21.	*Hordeum vulgare*	SUNESON	—	0	1	Anthers white, empty.
22.	*Lathyrus*	GREGORY	—	0	1	symbol ff.
		BATESON et al	—	0	1	,,
		PUNNETT	—	0	1	symbol b_2b_2.
23.	*Lycopersicum*	CRANE	—	0	1	Anatomical.
		CURRENCE	—	0	2	
		LARSON et al	—	0	1	John Baer mutant
		LESLEY et al	—	0	2(3)	Pollen fails to mature. Starch reserves profuse.
		RICK	—	0	1	Several nonallelic mutants
		ROEVER	—	0	1	
24.	*Malus*	GAGNIEU	—	0	4	p_1 to p_4. Meiosis regular except in few cases.
		HEILBORN	—			Genic. Combinations not known so far.
25.	*Medicago sativa*	CHILDERS	—			Genic. Many?

S. No.	Material	Author	Plasm	Gene dom.	Gene rec.	Remarks
26.	Nicotiana	BHATT et al	—	0	2	Little pollen produced but non-functional.
		RAEBER et al	—			Petaloidy of anthers. Genetic.
27.	Oenothera	EMERSON	—	0	1	Pollen failure after meiosis
		HARTE et al	—	0	2	symbol fr fr ster ster.
		LEWIS	—	2	0	
		SHULL	—	0	1	symbol ls ls.
28.	Oryza sativa	HSU	—	0	3	Among indica-japonica hybrids
		ISHIKAWA	—	0	1	Pollen abortive. No dehiscence.
		NAGAI	—	0	1	Awned-sterile mutant.
		RAMANUJAM	—	0	1	
29.	Origanum vulgare	LEWIS et al		2	0	Gynodioecious.
30.	Papaya	STOREY	—	0	1	Androecium suppressed.
31.	Paspalum	BURTON	—	0	1	Anthers develop occasionally.
32.	Petunia	WELZEL	—	0	1	Two nonallelic mutants, ms_1, &
		FRANKEL	+	0	0	ms_2. Linked with parv locus.
33.	Prunus communis	CRANE	—			Genic indicated.
34.	Prunus domesticus	DORSEY	—			
35.	Prunus persica	CONNORS	—			Results in self-sterility in some
		SCOTT et al	—	0	1	varieties.
36.	Ranunculus acris	MARSDEN-JONES et. al	—	0	1	Maleness recessive
		WHYTE	—	0	1	
37.	Raphanus sativus	TOKUMASU	—	0	1	Two more alleles.
38.	Ricinus communis	CLAASSEN et al	—	0	1	Modifiers indicated.
39.	Rosa wichuriana	NICOLAS				Floral teratism. Heritable.
40.	Rubus idaeus	CRANE et al	—	0	1	Sex suppression involved.
41.	Rubus odoratus	LEWIS	—	0	1	Pollen development affected.
42.	Satureia hortensis	CORRENS	+	0	0	Gyndioecious.
43.	Silene armeria	CORRENS	—	0	1	Pistillody.
44.	Solanum melongena	JASMIN	—	0	1	Anatomical.
45.	Solanum tuberosum	SALAMAN et al	—	1 or more	0	Inheritance complex.
		YOUNG				Indicates heritable nature.
46.	Sorghum	KAJJARI et al	—	0	1	Anthers empty at anthesis time
		KARPER et al	—	0	1	
		STEPHENS	—	0	1	
		HOLLAND	+	0	0	
		STEPHENS et al	+	0	1	
47.	Trifolium repens	OATWOOD	—			Heritable.
48.	Vitis	LMO				
49.	Zea mays	BEADLE	—	0	1	Several nonallelic mutants.
		EMERSON et al	—	0	1	
		EYSTER	—	0	1	
		HADJINOV	—	0	1	(cited after EMERSON et al)
		JOSEPHSON et al		0	0	
		PHIPPS	—	0	1	ts_{1-4}. Reports that at least 2 dominant and 3 recessives control sex expression.
		RHOADES	+	0	0	
		SCHWARTZ	+	1	1	

are known in onion and sugarbeets where the viability of pollen appears to be under gametophytic control.

From a number of investigations on dioecious, subdioecious and monoecious species interesting genetic data on the inheritance of sex have emerged. As a general feature, all of them supported the assumption of bisexual potentiality of every individual so that sex reversals or dioecism are simply due to a shift or switch in genetic balance towards either of the two sexes. JONES (1934) dealt with this problem in detail with particular reference to his findings on the genetic basis of floral expression in maize. A gene ba(barren stalk) when homozygous makes a corn plant staminate by eliminating the ears while another gene ts (tassel seed) when homozygous converts the tassel into a pistillate inflorescence, so that by proper manipulation, synthetic dioecious strains could be produced. Another modification called tunicate-ear involves a dominant factor. For a complete discussion of many cases with similar as well as other types of sex-deciding trigger mechanisms, the reader may refer to a recent review by WESTERGAARD (1958). In short, let M and F represent the two sex potencies, male and female, respectively. Let there be several loci for each of these sets, viz. M_1, M_2, ... ,M_k and F_1, F_2, ..., F_n. Then there are three kinds of changes possible, namely recessive mutation, dominant epistatic mutation, or disbalance resulting from anormal segregation and recombination of these component factors, which can interfere with the normal requisites for bisexual expression. The end result whether predominance of male or female sex, can vary depending upon the nature of specific factors involved in such an aberration. Theoretically, therefore, an array of different functional triggers may originate while the presence of heterogamety or an irregular mode of allosome behaviour characteristic of the species could further complicate the mechanisms of sex determination.

In castor SHIFRISS (1956) has presented an interesting scheme to explain the sex instability of this material. Accordingly, a gene F for monoecism controls a genetically stable series of sex variants ranging from female f to strongly males; in sex reversals an unstable nuclear factor suppresses F thereby causing delay in the gene impact inhibiting male potency. This nuclear factor can itself mutate into forms allowing release of suppressor action and the time of gene impact, in this case also a measure of dominance, depends on "the dose of the

TABLE 2. Male sterility among interspecific hybrids.

S.	Genus	Parental species		Author	Remarks
		♀	♂		
1.	*Aegilops*	*caudata*	*truncialis*	KIHARA et al	Role of cytoplasm indicated.
2.	*Aegilops*	*longissima*	*Aucheri*	KIHARA	F_1 only 0.3% pollen-fertile. Cytoplasmic.
3.	*Begonia*	*schmidtiana*	*acuminata*	VILLERTS	Genecytoplasmic. No pollen produced.
4.	*Capsella*	*orientalis*	*Haegeri*	SHULL	Pollen aborts.
5.	*Capsella*	*bursapastoris*	*Haegeri*	SHULL	
6.	*Cirsium*	*oleraceum*	*canum* or *palustre*	CORRENS	Cytoplasmic. Monoecy changes to dioecy.
7.	*Epilobium*	*hirsutum*	*luteum*	MICHAELIS	Genecytoplasmic.
8.	*Geranium*	*endressi*	*striatum*	SANSOME	Genecytoplasmic. Anthers petaloid or contabescent
9.	*Geranium*	*albiflorum*	*pratense*	SANSOME	
10.	*Hypericum*	*acutum*	*montanum*	NOACK	F_1 completely pollen sterile.
11.	*Linum*	*floccosum*	*usitatissimum*	GAIRDNER GAJEWSKI	Genecytoplasmic.
12.	*Mucuna*	*deeringianum*	*niveum* or *hassjov*	BELLING	Two factor pairs. K & L.
13.	*Musa*	*balbisiana*	*acuminata*	DODDS et al	Meiosis irregular. Inheritance inconclusive.
14.	*Nicotiana*	*nudicaulis*	*rustica*	CHRISTOFF	Anthers degenerate.
15.	*Nicotiana*	*debeneyi*	*tabacum*	CLAYTON	Genetycoplasmic. Originated in amphiploid backcrossed to *tabacum*.
16.	*Nicotiana*	*Langsdorffi*	*Sanderae*	EAST	Plasm factor and S allele interact.
17.	*Oenothera*	berteriana	odorata	SCHWEMMLE	Genic.
18.	*Origanum*	*majorana*	*vulgare*	APPL	In F_2 ♂or♀ gave only ♂ in crosses with ♂
19.	*Paspalum*	*urvillei*	*malacophyllum*	BURTON	Pollen aborts.
20.	*Solanum*	*rybinii*	*chacoense*	KOOPMANS	Incompatible combination of genom and plasm.
21.	*Solanum*	*chacoense*	*rybinii*	KOOPMANS	Morphologically different from its reciprocal.
22.	*Solanum*	*tuberosum*	*acaule*	LAMM	Role of cytoplasm or an interaction.
23.	*Streptocarp: s*	*comptonii*	*grandis*	OEHLKERS	Genic or genecytoplasmic.
24.	*Streptocarpus*	*wendlandii*	*Rexii*	OEHLKERS	Genic or genecytoplasmic.
25.	*Viola*	*arvensis*	*tricolor*	CLAUSEN	Two to several recessive factors.

suppressor, the potential tendency of ff and the nongenetic influences." Studies of LEWIS and CROWE (1952, 1956) in *Origanum* provided another example involving epistatic gene action. Here the male sterility (gynodioecy) is due to an interaction between two dominant factors. F causes anther suppression, H is its suppressor, the females are FFhh or Ffhh while double recessive is lethal.

In tomato LESLEY and LESLEY (1939) noted the occurrence of partially unfruitful plants in F_3 population from a trisomic F_2 individual. On microscopic examination of their pollen, some male steriles were detected. These authors found a digenic control for male sterility but a 7 : 1 ratio in backcrosses to male-steriles suggested to the authors that a third pair of ms genes may have also been involved.

GARBER (1922) analysed the genetic basis for the partial pollen sterility occurring in various breeding materials of oats. He found it to be inherited as a digenic recessive in Victory oats. From his analyses of rye population for pollen sterility, MÜNTZING (1946) classified as many as 50% of the total of 610 plants under observation as pollen-sterile. However there seems to be no correlation between the meiotic irregularities present in such populations and the amount of pollen abortion (PUTT, 1954). PUTT crossed low × high pollen sterile plants and found F_1 to have high pollen fertility. Moreover in F_2, F_3, F_4 generations from a cross between two inbred lines, a cytoplasmic control of pollen sterility quite independent from corresponding ovular fertility became evident. His results thus indicated a complex basis of inheritance involving both nuclear and plasmatic factors. LITTLE et al (1944) sampled a number of populations of onion varieties to show that there occurred even more complicated segregation of genotypes giving a complete array from homozygous to heterozygous ones.

In next section (Sect. VI) it will be seen that male sterile mutants exhibit an entire series of different modes of expression. CORRENS (1928) and many others have provided instances where disbalanced sex ratios, pistillody or anther suppression were governed by Mendelian factors. Pollen sterile mutants are often subject to cytological study of their irregular microsporogenesis or pollen mitoses. From genetic viewpoint this aspect can be attacked perhaps at all levels, viz. physiological, constitutional or cellular, depending on the nature of inhibiting factors as to whether enzymatic or substrate deficiency, or a catalytic shift in reactions.

B. *Cytoplasmic sterility.*

This type of male sterility is much less common except for the case of corn. (vide Table I). Plasm factor for the gynodioecious form was first conceived by v. WETTSTEIN (1924) and further studied elaborately by CORRENS (1928). Among crop plants, examples of cytoplasmic male sterility are known in flax (BATESON et al, 1921; and others), onion (MONOSMITH, 1926; JONES and CLARK, 1943; PETERSON and FOSKETT, 1953), maize (RHOADES, 1931, 1933; Jo-

SEPHSON and JENKINS, 1948; SCHWARTZ, 1951; BRIGGLE, 1953), sugarbeets (OWEN, 1945) and sorghum (STEPHENS and HOLLAND, 1954). In most cases it is found difficult to establish an unequivocal evidence of purely cytoplasmic inheritance. For the male-sterile maize RHOADES (1933) procured following evidence: "All of the progeny from crosses between male-fertile and male-sterile plants comprised of completely or partially pollen-sterile plants. There was no trans-mission of male-sterility through the pollen of partially sterile plants and it was possible to demonstrate that a systematic replacement of the entire genom present in the original sterile plants did not restore fertility. The nature of pollen parent had thus no apparent effect upon the expression of male sterility. The reciprocal crosses give altogether different result." In case of *Solanum chacoense* × *rybinii* hybrid, KOOPMANS (1951, 1955) provided the differential behaviour of reciprocal hybrids as more important evidence.

EDWARDSON (1956) has extensively reviewed the known cases of cytoplasmic and genecytoplasmic male sterility under several heads according to the source material whether an intergeneric, interspecific or an intraspecific hybrid or otherwise a pure species. In order to be able to detect cytoplasmic and genecytoplasmic modes of inheritance, we may go a step further, that is, it must be shown that the male sterility induced by the mother cytoplasm is dependent or independent of the genetic constitution of pollen parent. In other words, for a genecytoplasmic case, the nuclear factors, or the genome as a whole, would behave as plasmon-sensitive insofar they manifest different effects in different cytoplasm. Male sterility reported in *Aquilegia* hybrids (SKALINSKA, see EDWARDSON, 1956) is a case in instance. F_1 offspring of the cross *A. truncata* ♀ × *flabellata* ♂ showed a high percentage of pollen and ovules functional, whereas the reciprocal had nondehiscing anthers and 13–24% fertile ovules.

The controversy about the exact nature of plasm factor, whether it is a plasmagene or plasmon (see SIRKS, 1938 and MICHAELIS, 1954, for a general discussion) cannot be resolved by any of these methods. In corn, the mode of distribution of the male-sterile and normally fertile individuals with respect to the position of seed on cob did not seem to support the idea of discrete particles in cytoplasm that might segregate differentially to the male or female side prior to meiotic division. GABELMAN (1949, 1950) on the other hand, inferred from his study of the reproduction and distribution of this plasm factor that it consisted of certain unitary particles of which the presence of one to several in a microspore inhibits its development. According to him, the negative binomial distribution made up by a series of Poisson distributions closely parallels the behaviour of chromosomes. The occurrence of as many as dozen cytoplasmically male-sterile lines in maize arising independently seems to involve an autonomous cyto-plasmic component. It is now known that plastids which are highly autonomous in reproducibility, can be induced to mutate irreversibly through the action of a nuclear factor (for instance, iojap in maize; RHOADES, 1943). Interesting enough, later RHOADES (1950) could obtain a case of plastid mutation producing male sterility by a similar kind of gene action.

For a novel approach to the problem of establishing discrete nature of the plasmatic factors, FRANKEL (1956) made use of a cytoplasmic male-sterile line of *Petunia*. Scions of fertile line grafted on to male-sterile stocks survived slightly longer than the reciprocal grafts and the progeny of former consistently included both fertile and sterile individuals whether selfed or sibcrossed. Transmissibility of this character through grafting may be suggestive of the particulate nature of cytoplasmic determinants.

It may be well to note that cytoplasmic malesterility is often much more unstable than genic type. Perhaps this may also hold true of most other characters governed by cytoplasmic heredity. Under different environments, a male-sterile line might then show varying degrees of recovery in its fertility status. Furthermore the discovery of many fertility restorer genes in maize which apparently counteract in some way the effects of plasm factor, stimulated great interest in the genetics of this trait (JONES, 1951; JONES and MANGELSDORF, 1951; BRIGGLE, 1956; EDWARDSON, 1955; DUVICK, 1956; and many others). A possible interpretation of this restorer action could be to assume that purely genic or purely cytoplasmic cases are wellbalanced as counterparts which thus occasionally show up as an interaction phenomenon. In certain malesterile materials, however, BRIGGLE (1957) also observed some amount of selective fertilization taking place in plants bearing the cytoplasmic factor and in those hetero-zygous for the fertility restoring gene(or genes) from an inbred line.

C. *Genecytoplasmic male sterility.*

Classical studies on cytoplasmic inheritance in genus *Epilobium* have shed much light on this subject (MICHAELIS, 1933; MICHAELIS and MICHAELIS, 1948; many associates of MICHAELIS; see MICHAELIS, 1954). Among the F_2 and backcross generations of intraspecific crosses in *E. hirsutum*, involving the races Giessen or Vienna as the female parent, several male-sterile plants occurred. None of the F_1 populations showed differences in reciprocal crosses among five of these races. MICHAELIS interpreted the control of male sterility by an interaction between a recessive gene and the cytoplasm of Giessen or Vienna. Segregation with other hybrids suggested a cytoplasm-polygene complex. In an interspecific cross, *E. hirsutum* ♀ × *luteum* ♂, F_1 was completely pollen fertile whereas the reciprocal combination gave partially fertile offsprings. Repeated backcrossing to *hirsutum* parent as pollinator gave individuals with the *hirsutum* genome in *luteum* cytoplasm which showed progressive increase in the amount of pollen sterility. Here again the character was shown to be under gene-plasm interaction; another classical example of this type of male sterility is known in flax. Male-sterile plants occurred among the progeny of advanced generation crosses between a procumbent type and the common tall flax. They were characterized by reduced petals and more or less complete anther degeneration. Procumbent ♀ × tall ♂ gave only ♂ in F_1 and in F_2 3♂ : 1♀ (male-sterile) (BATESON and GAIRDNER, 1921). Further studies by CHITTENDEN (1927), CHITTENDEN and PELLEW (1927) and GAIRDNER (1929) determined a recessive gene m giving male sterility in presence of "procumbent" cytoplasm. GAJEWSKI (1937) held the opinion that those procumbent types were from another species *Linum floccosum*, so that it was actually a case involving interspecific crosses where as we find from many other instances, the genecytoplasmic type of male sterility is common.

In *Dactylis glomerata* MYERS (1946) reported male sterility inherited through an interaction between one to probably several dominant genes and the cytoplasm. From a segregating progeny, he obtained male-steriles when the seed parent was quadruplex, triplex or duplex for sterility factor, fertile if it was multiplex and segregating if simplex. Some male-steriles were recovered from intercrosses between two fertile F_1 plants known to be simplex so that absence of matrocliny may be inferred.

Inheritance of male sterility in sugarbeets has been worked out by OWEN (1945, 1948, 1952). In crosspollinated varieties he found it to be cytoplasmic while among their artificially inbred populations several genes were found to be influencing the cytoplasmic mechanism. Assuming for simplicity, two genes X and Z and two types of cytoplasm N and S for normal and sterile respectively are controlling male-sterile character, then Sxxzz represents male-sterile, SXxzz or SxxZz semisterile with ultimately no functional pollen and SXxZz semisterile with some good pollen. OWEN designated the genotypes Nxxzz, NX–zz and NxxZ– as types 0, I, II pollen parents in order to be sure to use only type 0 in hybrid seed program. Analyses of progenies from several different male-sterile plants crossed to the same set of pollinators further showed that the mode of gene-plasm interaction considerably varies from one parental combination to another (OLDEMEYER, 1957).

JOSEPHSON and JENKINS (1948) crossed an inbred strain of corn to several pollen parent lines and genetic study of the resulting progenies indicated that at least two genes plus the cytoplasmic contribution were governing the male-sterile character. JONES (1951) combined the plasm factor, the plasmagene according to him, and a nuclear factor (the chromogene) for male sterility in same plants to show their independent existence as well as action. Investigations by BRIGGLE (1956) on an incompletely sterile single-cross and its fertile reciprocal showed that besides these major genes, several minor factors are having influence upon the expressivity of this character. On the other hand, the wellknown 'Kys' type of male sterility in corn, first reported by SCHWARTZ (1951), is inherited through an interaction between three factors, namely a dominant gene for male sterility, a dominant suppressor of this factor (S^{Ga}) operating gametophytically and a specific cytoplasm (S). Thus individuals either lacking in Ms gene, carrying the S^{Ga} factor, or possessing N cytoplasm are all male-fertile.

Recently PARKEY (1957) has found that the pistillate expression of a line of castorbeans was governed by an interaction between a specific pistillate-inducing cytoplasm (S) and at least one dominant gene (P) so that backcrossing a female line to a male parent homozygous for P factor would satisfactorily maintain it so.

Earlier in connection with male sterility in genus *Epilobium*, it was mentioned that during successive backcrosses, as the addition of *E. hirsutum* genome in *E. luteum* cytoplasm progressed, the male-sterile expression increased until it was absolute. A comparable situation has been metwith in *Nicotiana* (CLAYTON, 1950), *Sorghum* (STEPHENS and HOLLAND, 1954), in *Aegilotricum* (FUKUSAWA, 1953, et seq.) and several others. From here it seems evident that male sterility occurring in hybrid materials frequently results from a general disharmony between the genome and cytoplasm of the two parents. Studies by OEHLKERS (1938, et seq.) on interspecific hybrids of *Streptocarpus* are enlightening in this connection. It was shown that the genome of *S. rexii* and other species belonging to Rexii-plasm group have female tendency predominant over male whereas it is viceversa for the *wendlandii* group. Thus accordingly F_1 plants with Wendlandii-plasm have androecium underdeveloped and this abnormality could be intensified by introducing two *rexii* genomes into Wendlandii-plasm. OEHLKERS could further establish *S. polyanthus* as an intermediate between these two groups with reference to the sex tendencies.

In *Geranium* SANSOME (1936) found a male-sterile plant in F_1 of a *Endressi* × *striatum* cross and by a study of backcrosses to *striatum* parent, he postulated a recessive factor of *striatum* controlling male sterility in combination with *Endressi* plasm. However, RAINIO's data on intersexuality in other *Geranium* hybrids, recently analysed by KOOPMANS (1952), point to the familiar genome-plasm interaction mechanism for male sterility, or precisely, as the case may be, the transformation of ovary into anthers.

VI. MODE OF EXPRESSION

The main significance of this section primarily resides with the fact that at least in a majority of the heritable cases of male sterility, the genetic factors act with remarkable precision as to the exact stage and mode of breakdown in normal processes of pollen development. On purely arbitrary grounds we may discuss the cytological aspects of this phenomenon under four separate divisions of the entire syndrome of developmental stages involved.

1). *The Stage of Undifferentiated Anther Tissues and of Sporocyte Formation.*

A number of instances are on record in literature which involve the degeneration of anther tissues during their differentiation (for instance, in potato, YOUNG, 1923; in *Cucurbita*, SHIFRISS, 1945). The presence of rudimentary anthers in the flowers of many monoecious and dioecious species may be assumed to result from abortion during such an early growth stage. In *Fragaria*, LONGLEY (1926) reported male-steriles having anthers turned into minute staminodia and warty anthers were observed in *Capsella* by SHULL (1927), in maize by BEADLE (1932b) and in sorghum by the present writer (JAIN, 1956). On their microscopic examination one would only find mass of dark, misshaped tissues appearing very unhealthy.

In many cases that show regular differentiation of the tissues of anther wall, archesporium and tapetal layers, the abortive process may initiate in any of these regions. In a male-sterile mutant of tomato RICK (1948) observed that instead of developing an outer wall of five or more layer thickness, the sporogenous cells were extending to the epidermis. The archesporium forms the precursor tissue of pollen mother cells and therefore any of its abnormalities would readily influence the normal course of microsporogenesis. OFFERIJNS (1938) obtained evidence in a male-sterile *Canna* showing that some early irregularity in the archesporial development inhibited the transition from pachytene to metaphase in abortive PMC's. NAKAMURA (1943) investigated a number of different types of male sterility in genus *Citrus* of which one occurring in Washington Navel orange involved an abnormal archesporial differentiation. Two other examples of breakdown at this stage were found in potato (YOUNG, 1923) and in sweetpeas (FABERGE, 1937).

Tapetum plays a vital nutritive role in normal pollen development. Its characteristic cataclysmic mode of development involves a phase of enlargement soon followed by a progressive degeneration during the meiotic division so that by the time of anthesis only vestigial remains are seen. Studies as those by RICK (1948) and CHILDERS (1952) on male sterility caused by tapetal abnormalities are enlightening. From a study of nine tomato mutants, RICK demonstrated that growth and functioning of tapetum could get retarded, accelerated or otherwise somehow deviated from the normal. The manifest failure of microsporogenesis varied as to the ultimate stage observed. CHILDERS found in some completely male-sterile plants of alfalfa that tapetum turned dense and swollen, the sporogenous cells much vacuolated in their cytoplasm, the time of onset of meiosis variable and the cell walls degenerative. Tapetum then disintegrated leaving anther locules full of oily droplets. He had also noted the lack of perfect adherence of the tapetum to the adjacent tissues.

WELZEL (1954) worked out the exact mode of pollen abortion in two male-sterile mutants of *Petunia hybrida*. In one of them, the abnormalities of cell division first showed up at diakinesis when the cytoplasm becoming shrunken, the bivalents did not spiralize any more and spindle inactivation resulted in noncongression of bivalents at the metaphase plate. The tapetum rapidly turned degenerative. In the case of another mutant, the microspores appeared to form normally but at this stage certain greasy substances passing from tapetum into them seemed to stay as droplets causing cessation of further development. In *Oenothera* a digenic case of male sterility was reported to show an abnormal tapetal segregation so that the mature pollen degenerated (HARTE and BISSINGER, 1952). Also the preliminary reports of GATES (1911) and OEHLKERS (1927) on male sterility in *Oenothera* and of CLAUSEN (1930) on a case in *Viola* hybrids suggested tapetal nonfunctioning as the immediate cause of pollen failure. In certain members of Gentianaceae that are normally devoid of any wellformed tapetum, the nutritive function is taken over by some of the spore mother cells which ultimately abort (GUERIN, 1926).

Cytological studies by MONOSMITH (1926) in onion, by CROWDER (1953) on *Festuca-Lolium* hybrids and of DELASTAING (1954) on *Salvia* species revealed clearly how some of these tapetal abnormalities bring about meiotic failure or if pollen does form, its degeneration. A definite gradient in the degree of abnormality going from periphery to the middle of locular cavity has been frequently observed; that is, cells nearest to the tapetal layer are less abnormal-looking than the farther ones. Significance of such a spatiotemporal interrelationship needs no overemphasis here. DARLINGTON and HAQUE (1955), for instance, showed convincingly that in the anthers of a male-sterile clone of *Allium ascalonicum*, two desynchronised groups of cells arose from a preceding timing upset in premeiotic divisions ultimately resulting in several different abortive processes.

In a male-sterile line of sugarbeets, ARTSCHWAGER (1947) observed irregular plasmodial development which he believed to result either from some kind of nutrient deficiency or a sudden release of metabolic waste from the vacuoles. SNOAD (1954) however attributes a similar abnormality occurring in his *Helianthemum* material to the restitution taking place during premeiotic mitoses caused in turn by spindle abnormalities. Another identical chain of breakdown events initiating at this stage is found in cases of syndiploidy, that is, the occurrence of binucleate PMC's. HOLDEN and MOTA (1956) reported in F_1 individuals of a cross *Avena barbata* $\times$ *A. strigosa* that these binucleate pollen mother cells suffered from nonsynchronisation with reference to the progress of meiosis in sister nuclei of which the peripheral one ultimately degenerated. They also reviewed a number of cases where the occurrence of such PMC's due to other reasons than restitution are known. It might be pointed out that many of those do not pertain directly to our discussion of the phenomenon of male sterility.

2). *Stage of Microsporogenesis.*

After the spore mother cells have normally differentiated, a series of breakdown processes can bring about such initial aberrations as the total suppression of cell division, a delay in its onset or its replacement by some mitosis-like division. In Veitchberry raised from the cross *Rubus rusticanus* $\times$ *idaeus*, CRANE and THOMAS (1949) found that anthers usually failed to develop, or when they did develop any far, microscopic exmination gave only isolated 'islands' of PMC's among the somatic tissues. BREEZE (1921) found in male-sterile potato plants that although PMC's formed normally, no meiosis took placc. The cytoplasm in them appeared to shrink away from the cellwall as if plasmolysis was occurring (YOUNG, 1923). In a tomato mutant studied by RICK (1948), the erratic orientation of sporogenous tissues led to its total collapse. JOHNSSON (1944) has investigated such instances in *Alopecurus mysuroides* more extensively and he found that with a marked delay in onset of meiosis, sporad stage was not reached and the division processes either stopped during first or the second division resulting in many irregularities. In *Viola* CLAUSEN (1930)had reported that PMC's of sterile anthers do not proceed beyond first division since the prophase was delayed and certain fibrillar attachments projected from those PMC's. Examples with 'mitotized' substitutes of meiotic divisions may be found in many apomicts (e.g. *Allium, Hieracium*).

In two haploid plants of *Phleum pratense*, LEVAN (1941) observed many large syncytes resulting from the fusion of several sporocytes with each other. These syncytes usually atrophied by the end of first meiotic division. To the above list of cases involving breakdown during an early developmental stage of the pollen mother cells, we may add those reported in *Oenothera* where the sporocytes failed to round off normally (GATES, 1911), in *Lathyrus* showing granulation of cytoplasm and abnormal contraction of chromosomes (FABERGÉ, 1937) and in *Viola* where unbalanced chromosome sets formed during some premeiotic division (CLAUSEN, 1930).

Deviations from normal meiosis are perhaps too numerous to allow mention of all of them. However to point out the more critical stages where the pollen abortion is frequently initiated, should be very useful for our discussion. Among them we may name the pachytene, first metaphase and anaphase, sporad, cytokinesis, microspore divisions and the mature pollen.

In male-sterile plants of *Hebe subalpina*, FRANKEL (1940) observed the degeneration of pollen mother cells at pachytene after the pairing was complete. In one of the male-sterile mutants of maize studied by the present writer, the PMC's aborted at leptotene as the nuclear material seemed to dissolve away leaving the anthers empty and nonexserted. RICK (1948) found a number of male-sterile tomato mutants showing irregularities during the first prophase and in two of them complete failure at this stage. Tapetal irregularities in them have been mentioned earlier.

In most instances of asynapsis or desynapsis the ovular fertility is simultaneously impaired. However, cases involving male sterility are known in *Zea* (BEADLE, 1930), *Hevea* (RAMAER, 1932) and *Alopecurus* (JOHNSSON, 1944). Among the possible reasons for asynapsis in last-named genus, JOHNSSON mentioned precocious chromosome division, nonhomologous associations, chiasma failure or deficient terminal affinity. Following an irregular anaphase separation, two unequal nuclei formed, one to several dicentric chromatid bridges appeared at dyad stage and cytokinesis took place precociously in relation to the chromosome cycle.

In *Kniphofia* MOFFETT (1932) reported spindle failure at various stages of first and second divisions but although this could account for the entire range of meiotic irregularities that were observed, it appeared that some genotypic reason brought about a high amount of sterility. In both *Hevea* and *Kniphofia*, the major breakdown of pollen development occurred after tetrad formation. From her studies on a male-sterile sugarcane plant, JANAKIAMMAL (1941) considered spindle activity to have considerable influence on the chromosome behaviour throughout during meiosis. UPCOTT's (1937) findings in a male-sterile mutant of sweetpeas are of especial interest. She reported the chromosomes at first metaphase being much shorter than normal and the chiasmata having largely terminalized. Also the timing unbalance between the chromosome cycle and the anther development starting at pachytene appeared to become acute by the end of first division.

A metaphase plate may sometimes suffer from lack of compactness or disturbed orientation of the chromosomes. DODDS and SIMMONDS (1946b) on the other hand observed in male-sterile plants of *Tripsacum laxum* that a sort of overcrowding at metaphase plate resulted in irregular disjunction at following anaphase.

LOWIG (1928) found pollen sterility in plants of *Iris pallida* to be associated with loss of chromatin during the first division. SCHNELL (1948) was inclined to attribute an identical irregularity in an interspecific hybrid of *Solanum* to the hybridity per se. In this case the metaphase pairing was seen to be retained much longer than normal duration.

At first anaphase such abnormalities as an inequal separation of chromosomes, occurrence of laggards or various difficulties of polar movement have been frequently encountered. In some male-sterile plants of potato, ELLISON (1936) reported the formation of restitution nuclei during first telophase which finally result in the formation of nonviable polyad pollen grains. However, generally in cases involving breakdown at one of these meiotic stages, the same individual might show various types of nonnormal behaviour so that it may seem that the governing mechanism is rather unspecific.

In most instances, these meiotic abnormalities reveal themselves in the form of degenerative tetrads, aberrant cytokinesis or the occurrence of microcytes. Besides the cytological causes, the tetrad stage is known to be critical one. In many *Citrus* species and their varieties, NAKAMURA (1943) recorded meiotic failure at this stage and according to him, nutritional factors of many kind had been involved. A rather common abnormality during this stage is the prolonged coherence of the four cells in a quartet until finally they abort, or proceed further as improperly differentiated (in *Vitis*, DORSEY, 1914; in potato, BREEZE, 1921; in tomato, RICK, 1948; in *Trillium*, KYONO, 1955). Failure of pollen development during or soonafter tetrad formation has been reported also in peach (ASAMI, 1927), rose hybrids (ERLANSON, 1931), maize (RHOADES, 1931) and in *Kniphofia* (MOFFETT, 1932).

Instances are known where either of the two meiotic divisions was completely lacking. However the second division is relatively more often found missing (FUKUDA, 1927; ARNASON, 1941; RAMAN, 1955). The irregularities of cytokinesis may as well be added to the list. BEADLE (1932a) in maize and RAMAN (1955) in jasmines provided examples where it was lacking altogether.

3). *Stage of Microspore Growth and Pollen Mitoses.*

The stage of transition of tetrads into young microspores is also critical one as may be deduced from the fact that many types of irregularities make appearance here. For instance, shortly after the component sporecells of a tetrad are released, their growth may retard or cease altogether so that failure to form an exine would bring about abortion (YOUNG, 1923, in potato; KNOWLTON, 1924, in peach; FRANKEL, 1940, in *Hebe traversii*). Several investigations have confirmed that this kind of failure was chiefly brought about by nutritive factors. In this connection the findings of COOPER (1952) are important to note. He had shown that a waxy material synthesized in the cytoplasm of tapetal cells was secreted into the locular cavity where it is used up by the microspores during their development.

VALLEAU (1918) in strawberry and DORSEY (1919) in plum found male sterility to result from the abortion of microspores during their growth and maturation divisions when a large vacuole was seen to appear in them. This kind of pollen failure in his *Citrus* material NAKAMURA (1943) designated as satsuma type. It was assumed to be involving both vegetative and generative nuclei in process of degeneration. In some of the pollen-sterile grape plants, DORSEY (1914, 1915) could clearly demonstrate that shortly after the first maturation division is complete, sterility assumed either of these two general forms: (a), Degeneration occurring only in the generative nucleus, or (b), degeneration involving both nuclei. As first signs of an abnormal behaviour of the microspores he observed shrinking of the nuclear membrane, conglomeration of the chromatin material and a rapid cessation of growth. In a similar manner, FUKASAWA (1956) showed in male-sterile *Aegilotricum* plants that microspores developing normally until about germpore formation, underwent degeneration initiating during or even little before the first mitotic division.

LACOUR (1949) obtained critical evidence to show that a proper differentiation of the microspore nuclei depends to a large extent on their polarity and orientation. In a clone of *Tradescantia* he found that errors in these factors gave rise to dwarf and nonfunctional pollen. It appears that certain qualitative differences in the cytoplasm surrounding the vegetative and generative nuclei play an important role in their normal development. As a result of erratic differentiation and timing upset, the vegetative nuclei underwent supernumerary divisions in the malesterile plants of *Lathyrus odoratus* (UPCOTT, 1937). It is interesting to note here that in an *Avena* hybrid showing the occurrence of binucleate pollen mother cells, HOLDEN and MOTA (1956) were able to establish a correlation between the cytoplasmic gradients observed in these anormal PMC's and the pollen grains at maturity.

4). *Stage of Mature Pollen.*

In the cases of pollen sterility that are unconnected with any observable meiotic abnormality or an anomalous maturation division of the microspores, the failure of pollen at maturity in all probability results from a genotypic cause. Such genic cases of postmeiotic pollen abortion or failure are known in *Oenothera* (OEHLKERS, 1927; HARTE and BISSINGER, 1952), in *Allium* (LEVAN, 1935), in *Nicotiana* (CHRISTOFF, 1938), in *Petunia* (WELZEL, 1954), in tomato (RICK, 1948 and later) and in maize (BEADLE, 1932b; and several others). It may be pointed out that these examples of specific pollen lethals are not at all identical with those involving numerical or structural causes as found in case of haploids, autotriploids or species hybrids. In fact these two types of sterility have been distinguished as haplontic and diplontic respectively although instances have been found which disallow such a classification without difficulty.

In the foregoing discussion of various modes of the cytological expression of male sterility, illustrations for each of the socalled critical stages in the pollen development were taken from many different species. It might be convincing to recapitulate a similar stepwise inhibition reported within the male sterile mutants of maize. In Table 3 a list of some wellknown cases of male-sterility in maize has been made with respect to the stage of abortion in the formation of pollen grains.

We have thus arrived at a stage when a hypothesis might be constructed to explain a genetic mechanism underlying a stepwise chain of phenotypic anomalies exemplified in above. Even an oversimplification may not be any less useful. Analogous to the scheme putforward by WESTERGAARD (1958) for the trigger mechanism of sex expression, it may be assumed that for each of those different steps in the differentiation of spore-mother cells, the meiotic divisions in PMC's and the formation of mature pollen, a single factor is responsible for the control of normal process and thus act in a series one after another. Biochemically one may also speak of one-gene-one enzyme mechanism to have been possibly involved. A male-sterile mutant would be the result of a socalled genetic block, the exact stage of its occurrence depending upon the specific function of the factor mutated. It is however difficult to advance our model any further until experimental information is forthcoming in support or otherwise.

TABLE 3. Genes for male sterility in maize.

Mutant	Symbol	Mode of expression	Author
antherless	at	Anthers rudimentary	HADJINOV (cited after EMERSON et al, 1935).
tassel-seed	ts_{1-4}	Terminal inflorescence predominantly pistillate	PHIPPS, 1928; EMERSON, 1928
warty anthers	wa	Sporogenous tissue degenerative in certain regions.	BEADLE, 1932b
male-sterile	ms	Nuclear material collapsing during leptotene	JAIN, 1956.
male-steriles	ms_8&ms_9	Fail during pachytene or at cytokinesis of first division.	BEADLE, 1932b
variable-steriles	va_{1-2}	Irregular division and distribution of cytoplasm. Amount of sterility varied.	BEADLE, 1931.
asynaptic	as	Asynapsis. Usually no pollen shed.	BEADLE, 1930.
sticky	st	Meiosis very irregular. Sterility usually complete.	BEADLE, 1932b
male-steriles	ms_4&ms_{10}	During cytokinesis after second division.	BEADLE, 1932a.
male-sterile	ms_1	Growth arrested soon after cytokinesis.	SINGLETON and JONES, 1930.
male-steriles	ms_3, ms_1&ms_{14}	Microspores abort before first maturation division	EYSTER, 1931; BEADLE, 1932b
polymitotic	po	Supernumerary divisions in microspores. No pollen shed.	BEADLE, 1931.
pollen-lethal	lp	Mature pollen degenerates.	RHOADES, 1950
vestigial-glume	Vg	Naked staminate flowers dry off prematurely.	SPRAGUE, 1939.
male-sterile	ms_6	Abundant pollen forms but anther tip pinched so that it does not dehisce.	BEADLE, 1932b

VII. INDUCTION EXPERIMENTS

Conscious attempts to induce male sterility have been only few although many experiments designed primarily for the study of effects of chemical substances and radiations on plants encountered pollen abortion as a conspicuous aftereffect. Insofar they might allow us to interpret various biochemical or physiological aspects of this phenomenon of male sterility, a reference would appear pertinent here.

1). *Maleic hydrazide and other growth substances.*

Table 4 below gives a summary of the main results obtained with maleic hydrazide and a few other growth chemicals. Success achieved with the use of maleic hydrazide for artificially inducing the pollen sterility appears encouraging. This chemical was ascribed its unique growth regulating properties for the first time by SCHOENE and HOFMANN (1949) whereas LEOPOLD and KLEIN (1951) discovered its antiauxin nature. It may be easily realized that, however, monoecious species like maize and a majority of cucurbits seem to be especially suitable material for such studies. Papers by WITTWER and HILLYER (1954) and by REHM (1952) deserve special mention in reference to the variable nature of the male-sterile expression in treated material. Besides, such problems immediately arise as those dealing with the proper growth stage of treatment, the length of period elapsing before recovery and the corresponding ovular fertility figures. Apparently, in order to evaluate the success of an induction experiment these and other associated factors should be kept in view.

WITTWER and HILLYER, using maleic hydrazide on plants of *Cucurbita pepo*, induced male sterility with several different methods of treatment. For instance, dipping or spraying an aqueous solution of 250–300 parts per million strength when the first leaf was expanding and repeating when 4–5 true leaves had developed, gave plants with

TABLE 4. Chemical inductilon of male sterility.

Chemical	Dosage	Species	Results	Author
1. Maleic Hydrazide	0.25%	Corn hybrids	No anther development	Naylor (1950)
2. ,,	0.1, 0.05 & 0.25%	Sweet corn	Successful	Denisen & Haber (1950)
3. ,,	250–300 ppm. or 100 ppm. five times	Cucumis and Cucurbita	No staminate flowers	Wittwer & Hillyer (1954)
4. ,,	2–8 lbs. per acre	Corn	Partially successful	Eskew & Willard (1950)
5. ,,	0.025%, 0.05, 0.1 & 0.15%	Corn hybrids	No effect	Warren & Dimmock (1954)
6. ,,	600 ppm.	Sweet corn	Female side also affected	Moore (1059)
7. ,,	250 & 500 ppm.	Watermelon	No effect	Rehm (1952)
8. ,,	500 ppm.	Tomato	Male sterility for two weeks	,,
9. ,,	0.025%	Corn	No anthers develop	Naylor (1950)
10. ,,	0.02–0.025%	Xanthium	Only photoperiodic response affected	,,
11. ,,	0.2, 0.4 & 0.8%	Tobacco	Inhibits flowering	,,
12. Tri-iodobenzoic acid	25,50 & 100 ppm.	Tomato	Male sterility for 8 weeks	Rehm (1952)
13. ,,	50 or 100 ppm. once or 10 ppm. several applications	Watermelon	Partially successful	,,
14. ,,	25 ppm.	cucumber	Sex ratio changes as male flowers are delayed	Wittwer & Hillyer (1954)
15. α-Naphthaline acetic acid	100 ppm.	,,		,,
6. ,,		,,	More female flowers produced	Laibach & Kribben (1949)
7. 2,4-Dichlorophenoxyacetic acid	10, 25 & 50 ppm.	Tomato	Inconclusive	Rehm (1952)
8. ,,	5 ppm. every week	Watermelon	Male sterility for one week	,,

usual number of female flowers while the androecia in male flower buds had aborted. They also found that repeated application of 100 ppm. for 4–5 times during the period of cotyledon expansion to the formation of a few leaves effectively suppressed the development of male flower buds. In a preliminary study using maleic hydrazide on wheat, the present writer showed the stage of treatment to be a critical factor. Application of 100 ppm. thrice or 250 ppm. only twice gave best results when the treatment was begun during the early tillering period whereas treatment at the time of flag leaf emergence affected gravely the ovular fertility as well. Furthermore, a similar dosage applied to tomato resulted in extreme inhibition of flowering Thus response to the chemical varied greatly from species to species. In fact, McIlrath (1953) could demonstrate such differences within a few varieties of sorghum. From here it follows obviously that extensive research alone might yield a consistent and precise method of inducing male sterility. To be useful in field crops, the method should be easy, economical and also reproducible. Fortunately the effective concentrations of these substances are enough low so that toxicity does not seem to pose a problem. In addition, at these concentrations chromosomes are not any seriously affected which must be an additional advantage in this kind of work.

From these few reports on the successful use of growth regulating substances in induction of male sterility, a paradoxical situation has to be considered. How an auxin (viz. α-NAA) and an antiauxin (viz. MH) act to bring about the same result? Before we might attempt to speculate on this, a preliminary knowledge of some of the primary activities in anthers during the gametophyte development would seem essential. A brief chart may be given to present them here.

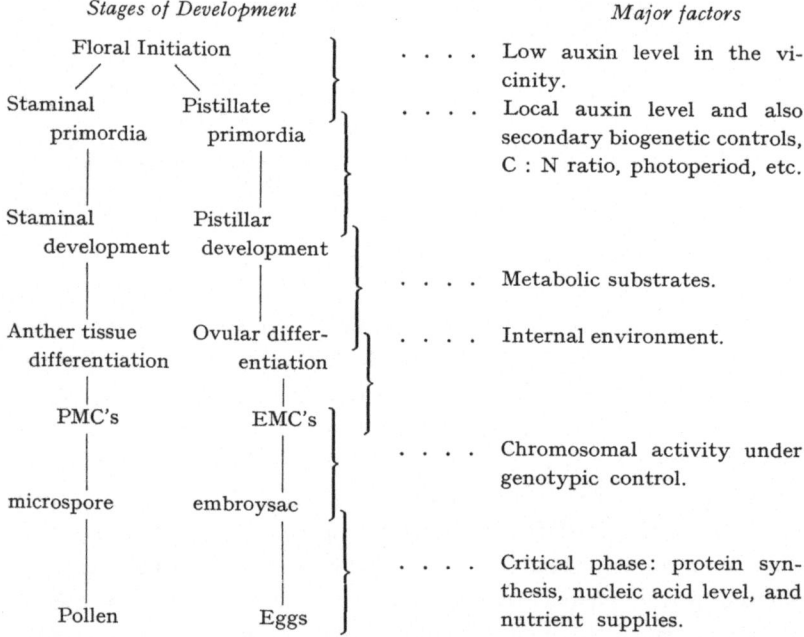

Stages of Development   *Major factors*

Floral Initiation   .... Low auxin level in the vicinity.

Staminal   Pistillate .... Local auxin level and also
primordia   primordia   secondary biogenetic controls,
                                 C : N ratio, photoperiod, etc.

Staminal   Pistillar
development   development
                            .... Metabolic substrates.

Anther tissue   Ovular differ- .... Internal environment.
differentiation entiation

PMC's   EMC's
                            .... Chromosomal activity under
                                 genotypic control.

microspore   embroysac

                            .... Critical phase: protein synthesis, nucleic acid level, and
Pollen   Eggs   nutrient supplies.

Any upset in one of the major factors listed above would provide a pathway to lead to male sterility, considering only the male side here. In light of this information, certain findings from physiological studies with maleic hydrazide are of direct interest. GRELAUCH (1951) and MᶜILRATH (1950) reported that plants treated with this chemical accumulate a good amount of sucrose in their leaves, in turn causing alterations in the carbon-nitrogen ratio of these plants. An important idea postulating the hormonal regulation of the nucleic acid metabolism was first conceived by SKOOG (1953) who could show a correlation between auxin changes and the proportionate changes in the ribonucleic acid content. In his opinion, "often very striking similarity in the growth responses elicited by each of these agents and by the abnormal growth relations or genetic factors would seem to be a predictable phenomenon worthy of further study". The important role of RNA/DNA ratio in normal pollen development has been severally indicated. In *Rhoeo* anthers, PAINTER (1943) found the cytoplasm of both tapetum and sporogenous tissues rich in RNA which changes into DNA as the pollen mother cells prepare for the meiotic divisions. Microspores, on the other hand, accumulate RNA in their cytoplasm as they grow into mature pollen grains.

In addition to these two effects of maleic hydrazide, namely changes in C : N ratio and in nucleic acid level, ISENBERG et al (1951) reported it to cause partial inactivation of respiratory dehydrogenases and LIVINGSTON et al (1954) its effect on free aminoacid content in sugarbeets.[1] The significance of these findings in explaining away the mode of action in inducing male sterility seems to be apparent.

In a few instances the observed irregularity of pollen development is found to be visibly associated with the occurrence of chromatin droplets in cytoplasm which when excluded, result in nucleic acid deficiency. (LIMA-DE-FARIA, 1947; NIELSEN, 1955). FUKASAWA (1954) analysed free amino acids in normal and sterile anthers of male-sterile emmer wheat plants by use of the paper chromatograms and found the latter to have an excess of asparagine alongwith a deficit in proline. It was not ascertained, however, that whether these differences were anyway related to a nutritive factor. From his studies with excised anther cultures and autoradiographs, TAYLOR (1950, 1953) has shown a critical phase at a corresponding stage to that analysed by FUKASAWA. With paper chromatography again EDWARDSON (1955)

---

[1] The highly specific nature of maleic hydrazide reactivity was shown by LOVELESS (1952) who used its analogues to demonstrate that its characteristic 3 : 6 pyridazine diôl structure can not be modified without a considerable loss in the activity of this chemical.

found that in a cytoplasmic male-sterile corn line, tassels possess a ninhydrin-positive material that is absent in normal fertile ones. In a similar material JONES et al (1957) reported precocious accumulation of alanine and other aminoacids in anthers.

Thus we already have indirect evidence on the nature of biochemical pathways possibly involved in pollen abortion brought about by treating with maleic hydrazide. On the female side such data are not available so far so that the differential effect on male side alone in these experiments remains a moot question. It is hoped that initial stimulus from these beginnings in biochemical direction would induce many investigations in near future.

2). *Irradiation.*

Since pollen sterility may be rather easily recorded, it is not at all surprising that some cases of radiation-induced male-sterile mutants have been readily detected. In tomato LINDSTROM (1933) obtained a case induced by radium treatment. From his studies on the X-ray-induced hereditary variations in *Triticum monococcum*, SMITH (1938, 1939) reported eleven cases of male sterility showing various abnormalities during microsporogenesis in some and the failure of mature pollen in others. He found all of them to be inherited as monogenic recessives. KRISHNASWAMI and AYYANGAR (1942) obtained likewise a recessive case of X-ray induced male sterility in *Pennisetum typhoides.* MORRIS (1952) with the use of thermal neutrons induced partial pollen sterility in corn and cytologically he detected several types of chromosomal irregularities in these individuals. Another paper reporting success by thermal neutrons is by YOST et al (1953) in *Datura.*

MOH and NILAN (1953), on the other hand, obtained a case of carpelloidy induced by atomic bomb radiations on barley seeds. In course of an experiment with X-rays and P^{32}, LESLEY and LESLEY (1955) found male sterility in tomato plants to occur during the R_1, R_2 and R_3 generations showing in the form of unfruitfulness. This had involved certain structural and other kinds of meiotic abnormalities in chromosome behaviour. It would appear from these experiments that pollen side is probably much more sensitive to radiation injury than the ovular side. There are, moreover, some reasons to suggest that a part of this differential sensitivity of pollen or its precursors is genotypically determined..

3). *Environmental modifications.*

Recently HESLOP-HARRISON (1957) published an excellent review of many studies on artificial induction of male sterility by altering such factors as nutrient supply, photoperiodic exposure and temperature conditions, or simply by mutilation, grafting and so on. For sake of brevity, therefore, only the main conclusions from these experiments need be included in below.

(i). Mineral nutrition. Moist soils containing a relatively high level of available nitrogen promote femaleness whereas dry soils low in nitrogen seem to induce maleness. While using such expressions as maleness or femaleness, it is assumed that different types of male sterility as described in this paper are kept in view. The careful experiments of SCHAFFNER (1929 and earlier), HOWLETT (1936). MININA (1938) and of HALL (1949) provided evidence to support these conclusions.

Working with cucumber, MININA compared the sexual expression in plants receiving a complete quota of mineral nutrients at the beginning of experiment with those given periodic doses. In latter series the sex ratio was found to be 39 males to only 6 females among first order flowers and 7 to 7 in second order. The corresponding ratios in controls were 58 : 4 and 15 : 4 respectively. Thus intermittent supply seemed to cause 'female sexualization'. HALL obtained an even more clear evidence for the fact that liberal nitrogen supply increases the proportion of pistillate flowers to the total. It should be added here that most of these experiments had a requirement of short day conditions in order to be successful in inducing male sterility.

(ii). Carbohydrate status. KRAUS and KRAYBILL (1918) were probably first to emphasize upon the role of carbon-nitrogen ratio in floral initiation. Subsequent to this publication, it was established that a low C : N ratio during the period of floral morphogenesis causes arrest of pollen development (WATTS, 1931), or some other form of male sex suppression. (STOUT, 1923; DARROW, 1927; JARETZKY, 1927). Earlier it was mentioned that maleic hydrazide may be acting by influencing this factor to give successful induction of male sterility. In an interesting paper on the male-sterile case studied by FUKASAWA et al (1957), it has been made evident that with a better supply of sugars the pollen development progressed better. In certain instances of pollen failure presumably due to this reason, spraying a sugar solution on pollen sterile plants markedly improved the fertility (SWAMINATHAN, 1952).

(iii). Light. This is a potent factor in the determination of sex tendencies either through the control of carbohydrate metabolism or as such by way of photoperiodic responses. In a number of both dioecious and monoecious short day species, continuous short day exposures shifted the sex balance towards predominance of femaleness (male sterility in our sense). In support, the results of TOURNOIS (1911) in *Humulus*, RICHEY and SPRAGUE (1932) in maize, NAYLOR (1941) in *Xanthium* and of THOMPSON (1955) in spinach merit special mention. (cf. HESLOP-HARRISON's review). It has been carefully pointed out by HESLOP-HARRISON (1957) that "no confusion must be allowed between the effect of different photoperiodic regimes on sexual balance in monoecious and hermaphrodite plants and the phenomenon of complete or partial flower abortion found under some conditions of "borderline" photoperiodic induction".

(iv). Temperature. It has been very well established that at least in many monoecious species low temperatures during the early growth stages promote female sex and depress maleness. Evidence from investigations on the response of hemp and spinach in this respect is elucidative. On the other hand, MOL (1933) and STOW (1930) working with hyacinths obtained 'embryosac-like' giant pollen grains in the partially petaloid anthers of plants raised from bulbs having been subjected to high temperatures. These pollen grains might occasionally function as true embryosacs which indicates the versatility of some temperature-induced remote change in the sex-deciding trigger. NAKAMURA (1936) was able to demonstrate in many *Citrus* species that temperature conditions prevalent during the flowering stages differentially act on microsporogenesis. The results of THOMPSON (1955) represent the induced male dominance at higher temperatures due to some substantial shift within potential females.

(v). Experiments dealing with environmental modifications of sex expression, in addition to the important factors mentioned above, might involve the effects of mutilation, pruning, organic chemical (for instance phenyl mercurials, McFARLANE, 1950) and especially animal hormones and other kinds of manipulations of plants or their parts. In a few cases, the presence of a virus disease might affect the pollen development more than that of ovules, thereby resulting in male-sterile expression.

Although it may seem difficult to appraise many of these preliminary findings on the mechanism of sex determination or development
it may be hoped that future research would make it possible to devise
suitable techniques for the artificial induction of male sterility that
may have some value of theoretical as well as practical magnitude.

VIII. UTILIZATION IN PLANT BREEDING

It needs no overemphasizing that the use of male sterility in commercial production of the hybrid seed offers a promising tool in the hands of plant breeder. An initial wave of enthusiasm starting from the success reported in onion and sorghum, stimulated several attempts in other crops as well and in fact, successful projects are already underway in maize, tomato, barley and sugarbeets. In Germany a commercial seed firm obtained patent rights for the production of hybrid seed of sugarbeets using male sterility (DORST, 1952). The potent usefulness of such methods has been also indicated for many other crops including cabbage (DETJEN, 1927), guyale (POWERS and GARDNER, 1945), cucurbits (SCOTT and RINER, 1946), castor (CLAASSEN and HOFFMAN, 1950), tobacco (CLAYTON, 1951), cotton (LODEN and RICHMOND, 1951) and carrot (WELCH and GRIMBALL, 1947).

KLEY (1954) summarized a few more important general principles involved in its practical utilization on commercial scale. For sake of continuity some of them may be restated here.

Male sterility is most easily applied in crossfertilized plants and more so when it is cytoplasmic. Thus when all mother plants possess the sterile (S) cytoplasm, on growing the pollen parent lines with it, natural crossing would yield practically centpercent hybrid seed on the former without any need of emasculation or handpollination. Also in crops like beets or onion that are grown for their vegetative parts, the sterility of F_1 generation presents no problem.

For the dihybrid method in corn only half of the lines to be used as seed parent need to possess the plasm factor for male sterility. Two single crosses, respectively male-fertile and male-sterile, are obtained from sterile × fertile inbreds and fertile × fertile combination. These single crosses are then interplanted together to produce the double cross seed. However, since the plants raised from dihybrid seed would include some male-steriles, sufficient pollen sources need to be provided or otherwise a method to restore fertility has to be employed. A number of restorer factors have been worked out genetically and their application is soon expected to be imminent.

STEPHENS et al (1952) reported male sterility in the Day variety of sorghum and they putforth a plan of producing threeway crosses according to which three stocks are to be maintained. Stock A segregating for fertile and sterile plants in about 1 : 1 ratio are derived from seeds on the male-sterile plants each year. Stock B as pollen parent is sown alongwith A and single cross A × B interplanted with another normal fertile stock C. It seems that windpollination is enough effective as not to warrant need of handpollinations. Later, STEPHENS and HOLLAND (1954) have obtained male-sterile character governed by an interaction between the milo cytoplasm and kafir nuclear factors. With this, parent stocks are easier to develop and maintain and also only two instead of three isolation blocks would be required since the block necessary for maintaining the genetic male-sterile stock in heterozygous condition is rendered superfluous.

In selffertilized species it requires additional effort to maintain the male sterile lines in homozygous condition by handpollinations and in fact, only in a few horticultural crops the method has been a practical proposition. In tomato an interesting mutant has occurred in John Baer variety in which although abundant pollen forms, the anthers fail to dehisce on account of germ pore being absent. Since this mutant can be selfed by handpollination, it is easily maintained as a pure line.

It may be readily realized that in most cases certain general as well as specific problems have arisen to largely hamper a rapid progress in this field. It seems in order that some of them are discussed here.

(i). Identification of male-sterile plants. In order to succeed in a large scale survey of breeding material for male sterility or in producing large numbers of male-sterile segregates of any mutant, ease of their identification would be an important consideration. Cases involving the necessity of a microscopical examination of pollen or of even more tedious procedures like tests of pollen germinability or actual pollinations would be out of question, or at least present a great task. Some easy method of field identification should be desirable. In tomato, potato and many horticultural crops, male steriles are often conspicuous by their unfruitfulness if best growing conditions had prevailed. By that time it is usually late in the season for making controlled pollinations. In most male-sterile mutants of tomato studied by RICK (1948), identification was found possible by simply prodding the anthers for their contents or by use of a hand lens. In maize generally the sterile tassels appear very different as the anthers do not exsert from them. Also the empty anthers look shrivelled and feel flabby to the touch. In SUNESON's (1940) barley mutant, the loose open appearance of the sterile spikelets affords a convenient means of identifying male-sterile individuals. Often various discernible differences in the size, shape or the colour of anthers or certain marker genes may be found helpful. The close linkage of a male-sterile factor in maize with the yellow endosperm locus (SINGLETON and JONES, 1930), with awnless-awned and green-virescent loci in sorghum (STEPHENS and QUINBY, 1945), with potatoleaf and green stem characters in tomato (CURRENVE, 1944) and with seedling abnormality in limabeans (ALLARD, 1953) have provided positive methods of roguing the partially fertile segregates too, if any.

(ii). Natural crosspollination. This is another important practical consideration. It was mentioned above that in many cases interplanting the fertile and the male-sterile lines is all that is needed to get hybrid seed from the latter. RIDDLE and SUNESON (1944) had to perform masspolinations in barley. In case of tomato, RICK (1945, 1947) considered the natural crossing as welcome addition to the handpollinations in case the whole planting was adequately isolated from other stocks. Use of a recessive seedling character could be further helpful in eliminating the occasional selfs (LARSON and PAUR, 1948).

In crossfertilized species, on the other hand, the amount of self-pollination has to be investigated. Any high degree of selfing should be avoided. In onion, however, CLARKE and POLLARD (1949) found it to be only 4% which does not seem a serious problem at this level but as some plants produced even more selfed seeds, these authors suggested the need of selecting for high degrees of male sterility among the genotype mixtures at hand.

(iii). Stability of its expression. Use of male sterility in maize was discredited for a considerable time since first discovered mainly for the reason of its instability. In fact, the cytoplasmic type is often more unstable and variable in the extent of sterility. Studies by JOSEPHSON and JENKINS (1948) on the influence of planting dates and other environmental variables showed that intermediate planting season contributed toward a partial recovery of pollen fertility. Also ROGERS and EDWARDSON (1952) observed that the Texas cytoplasmic male-sterile lines lose expressivity of this character in some fields. It is therefore felt necessary that adequate testing of the response of any given line under a variety of climatic and cultural circumstances must be the first step in its utilization. For similar reasons a constant inspection of all the male-sterile stocks or their hybrids is essential in order to maintain the genetic purity in such a material. Vegetative propagation such as the use of cuttings in tomato makes this very easy.

(iv). Transfer of male sterility to other varieties. The ease of its transfer from the original source to the varieties more desirable from different viewpoints largely depends upon the mode of its inheritance. In onions, for instance, male sterile character has been readily transferred from the Italian Red-13 source to many other varieties where it is maintained by repeated backcrossing to the normal fertile line. (JONES and CLARKE, 1943; JONES and DAVIS, 1944). Such transfers were found essential for obtaining the parental combinations with highest combining ability. In this case the backcross seed is used to perpetuate the line as well as for producing male-sterile parents to be employed in hybrid seed production. KLEY (1954) has outlined a promising scheme by which only a limited number of backcrossing generations followed by one selfing generation give a genotype differing from the original parents in male sterility alone. In maize it was soon realized that the fertile plants whose pollen is to be used in backcrosses, must be selected for the complete absence of any restorer factor in them (GABELMAN, 1949; JONES and MANGELSDORF, 1951). This could be handled by using the method of paired progeny selection in which the pollen parent is selfed and grown alongwith the sterile backcross progenies.

Obviously, it is more effective to look for the male-sterile stocks within an original material rather than to transfer from others. It appears from success reported for the case of tomato and maize that a large scale search is often fruitful in this context.

(v). Restoration of fertility. Use of cytoplasmic male sterility in crops grown for seed raises the problem of restoring the fertility of hybrid generation because it would not then produce any seed. Incorporation of some restorer genes in the inbred used as pollen parent would be necessary. In maize the genetics of restoration problem has been a subject of widespread interest during past few years. Many workers have found a dominant factor to be controlling the fertility restoring ability of certain specific stocks. (JONES, 1951; JONES and MANGELSDORF, 1951; EDWARDSON, 1955; THOMAS and JOHNSON, 1956). The wellknown Kys type of male sterility was formerly considered useful solution to this problem since the fertility of double cross hybrid gets recovered itself while using normal inbred lines or their single crosses as pollinators (JUNGENHEIMER, 1951; BAUMAN, 1953).

DUVICK (1956) carried out an extensive test for allelic relationships among the genes for fertility restoration in five different FR (restorable) inbred lines. Three main conclusions could be drawn from it:
"a. All five lines have same allele of the major factor for restoring fertility in the strain WF9.
 b. Two or more inbreds equally sterile in T cytoplasm may not necessarily be isogenic for all FR lines.
 c. The environmental influence is often too great to allow reckoning of the different numbers of fertility restorer factors."

DUVICK was careful to point that all segregating populations should be grown in both N (normal) and S (sterile) cytoplasm in order to ascertain that one is not confusing the two separate modes of inheritance.

To make use of these factors in breeding programs, the fertile single cross parent should be made homozygous for one or more dominant alleles restoring fertility in the plasm carried by other parent singlecross. In order to prevent any loss of these factors during the process, each year testcrosses should be made with the available male sterile lines.

Lastly it may not be amiss to mention here that besides its utilization for hybrid seed production on a commercial scale, male sterility can be employed in many different types of experimental crossing work on almost identical lines. For instance, in sugarcane many biparental crossing and polycross projects are beginning to utilize this character (HAYES et al, 1955) and so also in barley for producing synthetic hybrids (SUNESON, 1945). ALLARD (1953) pointed out some such possibilities for the case of limabeans. Several other uses too have been known. In barley, SUNESON and HOUSTON (1942) used male-sterile plants for mass inoculations with a culture of pathogen transmissible through floral infection. In tobacco, to eliminate seed production topping is practised but this was done by CLAYTON (1950) using male-sterile character which makes it inexpensive. Mention should be made of its use in making estimates of the natural rate of crosspollination in any given material (RICK, 1949, in tomato; BHATT and KRISNAMOORTHI, 1956, in tobacco). In a recent paper by PERSSON and RAPPAPORT (1958), it was shown that the testing procedures for effectiveness of a chemical treatment on the induction of parthenocarpic fruitset were refinable by the use of male sterility. These workers investigated the effect of gibberellin in tomato, showing a positive increase in parthenocarpic fruitset with this chemical.

IX. EVOLUTIONARY SIGNIFICANCE

In earlier sections, few points were made on the modes of origin and occurrence of the phenomenon of male sterility in flowering plants. Theoretically at least it has manifold evolutionary importance. A study of distribution patterns of the diverse sexual forms among both monomorphous and polymorphous groups is expected to throw some light on the intermediate steps in the evolution of dioecism from hermaphroditism if one might assume that dioecy originated from the bisexual ancestors through mutation and natural selection. In general, the students of sex determination problem have tended to favour the hypothesis of a trigger mechanism governing the shifts in sex tendencies of a potentially bisexual individual, that is, naively speaking, primordia in floral regions destined to develop both sexes are directed towards either sex. The synthetic dioecious strains obtained by JO-NES (1934) and others in maize show with exactitude a genetic basis of the origin of dioecy (LEWIS, 1940, 1942). On the contrary, the trigger found in case of *Melandrium* seems to illustrate the retrogession from dioecism to bisexuality. WESTERGAARD (1958) has critically examined both of these cases on a comparative basis to show their relative efficiency and stability. It may be pointed out, however, that controversy about whether dioecy or hermaphroditism is the primitive condition loses meaning if a theory of the polyphyletic origin of Angiosperms was accepted.

Male-sterile mutants have been recurrently occurring among natural populations and in fact some of them seem to have established among the breeding materials of several species. FISHER (1930) has considered in his pioneering book The Genetical Theory of Natural Selection incidence of such mutants and chance of survival under the evolution of dominance. The stochastic changes involved in the fixation or otherwise elimination of such rare genic changes in populations have been of interest to population geneticists. In a brief paper, KLEY (1955) derived formulae to show the relationships between the frequencies of natural occurrence of genic and cytoplasmic male sterility in self- and crossfertilized species, their relative fertility and their mutation rates. Accordingly, the expected percentages of male-sterile plants in case of cytoplasmic, monofactorial dominant and monofactorial recessive are $p/1-q$, $2p/2-q$ and $p/2-q\%$ respectively, where p is the probability of its occurrence and q is the relative fertility expressed as the ratio between the average seed set per plant by male-sterile and normal individuals. Thus cytoplasmic type is expected to be most frequent in allogamous materials. CORRENS (1928) was first to draw attention to the prevalence of cytoplasmic male sterility among the gyndioecious species.

MATHER (1940) contended on purely theoretical grounds that gyndioecy as such provides a far more superior outbreeding mechanism than unisexuality, incompatibility or likewise any other contrivance to aid outbreeding for that matter. It should be pointed out in passing that gyndioecism as an outbreeding mechanism differs from others in that two mating types, that is females and hermaphrodites, contribute unequal numbers of genes to the next generation. According to MATHER, the separation of two sexes is more for providing panmixia rather than a means to ensure perfect gametic differentiation as was believed by earlier workers. LEWIS (1941, 1942) extended MATHER's ideas by showing mathematically that such a cytoplasmic setup as that occurs in gyndioecious species was an ideal and sensitive means of adjusting the amount of outcrossing to the hybridity optimum of a population, whereas genic type lacks in both of these requirements. In *Origanum vulgare*, LEWIS and CROWE (1956) provided supporting evidence for a slightly higher fitness of the females being the controlling factor of the sex ratio at an equilibrium. It was postulated that gyndioecy in this species has evolved from an incompatibility system with sporophytic pollen control and an inhibition of the fertilization in the embryosac.

It may be pointed out that in comparison with incompatibility system, at least certain types of male sterility possess an advantage of avoiding gametic waste of large measures. This and other related concepts had their strongest proponent in DARWIN who argued for the natural selection being a driving force behind the origin of outbreeding systems.

Genecytoplasmic male sterility as illustrated best in the genera *Epilobium* and *Zea*, has been considered to be an important source of the isolating mechanisms that are essential for the origin of new entities (STEBBINS, 1958). By a segregation of appropriate combinations of nuclear and plasm factors in a single population, such barriers may arise rather readily.

ACKNOWLEDGEMENT

The writer wishes to express his sincere thanks to Mrs. Dr. A. KOOPMANS and to Dr. M. S. SWAMINATHAN for their many helpful suggestions.

REFERENCES

ALLARD, R. W., 1953. A gene in limabeans pleiotropically affecting male sterility and seedling abnormality. Proc. Amer. Soc. Hort. Sci. 61: 467–71.

ALLEN, C. E., 1940. The genotypic basis of sex expression in Angiosperms. Bot. Rev. 6: 277–300.

APPL, J., 1929. Weitere Mitteilungen über die Aufspaltung eines Bastards zwischen *Origanum majorana* L. ♀ und *Origanum vulgare* L. ♂ in der F_2 und F_3 generation. Genetica 11: 519–558.

ARMSTRONG, J. M. and W. J. WHITE, 1935. Factors influencing seedsetting in alfalfa. J. Agric. Res. 25: 161–79.

ARNASON, T. J., 1941. Sterility in potatoes. Canad. J. Res. Sect. C Bot. Sci. 19: 145–55.

ARTSCHWAGER, E., 1947. Pollen degeneration in male-sterile sugarbeets with special reference to tapetal plasmodium. J. Agric. Res. 75: 191–97.

ASAMI, V., 1927. Pollen abortion in the Shanghai peach. J. Sci. Agr. Soc. 297–364.

ATWOOD, S. S., 1944. Oppositional alleles in natural populations of *Trifolium repens*. Genetics 29: 428–35.

AYYANGAR, G. N. R., 1931. *Eleusine coracana*. Part III. Sterility. Ind. J. Agric. Sci. 1: 554–62.

AYYANGAR, G. N. R. and V. L. SRINIVASARAO, 1941. Studies in Barnyard millet (*Echinocloa colone*). Madras Agric. J. 29: 3–12.

BARHAM, W. S. and H. M. MUNGER, 1950. The stability of male sterility in onions. Proc. Amer. Soc. Hort. Sci. 56: 401–09.

BATESON, W. and A. E. GAIRDNER, 1921. Male sterility in flax subject to two types of segregation. J. Genetics II: 269–76.

BATESON, W., E. R. SAUNDERS and R. C. PUNNETT, 1908. Male sterility in *Lathyrus odoratus*. Rep. Evol. Comm. Roy. Soc. Lond. 4: p. 16.

BAUMAN, L. F., 1953. Results of further inheritance studies on the Kys type of cytoplasmic male sterility in corn. Agron. Abst. Amer. Soc. Agron.

BAUR, E., 1924. Untersuchungen über das Wesen, die Entstehung und die Vererbung von Rassenunterscheiden bei *Antirrhinum majus*. Biblioth. Genet. Lpz. 4: 1–170.

BEADLE, G. W., 1930. Genetical and cytological studies of Mendelian asynapsis in *Zea mays*. Cornell Univ. Agric. Exp. Sta. Mem. 129.

BEADLE, G. W., 1931. A gene in maize for supernumerary cell divisions following meiosis. Cornell Univ. Agric. Exp. Sta. Mem. 135.

BEADLE, G. W., 1932a. A gene in *Zea* for failure of cytokinesis. Cytologia 3: 142–55.

BEADLE, G. W., 1932b. Genes in maize for pollen sterility. Genetics 17: 413–31.

BEASLEY, J. O., 1942. Meiotic chromosome behaviour in species, species hybrids haploids and induced polyploids in *Gossypium*. Genetics 27: 25–54.

BELLING, J., 1914. The mode of inheritance of semisterility in the offspring of certain hybrid plants. Zschr. ind. Abst. Vererb. 12: 303–41.

BHADURI, P. N., 1942. Cytological analysis of structural hybridity in *Rhoeo discolor* Hance. J. Genetics 44: 73–85.

BHAT, N. R. and T. KRISHNAMOORTHI, 1956. A male sterile mutant in *Nucotiana tabacum*. Curr. Sci. Bang. 25: 297–99.

BLAKESLEE, A. F. and J. L. CARTLEDGE, 1927. Sterility of pollen in *Datura*. Mem. Hort. Soc. New York 3: 305–12.

BLEDSOE, R. P., 1929. Multiple kernels in wheat-rye hybrids. J. Hered. 20: 137–42.

BREEZE, M. S. G., 1921. Degeneration of anthers of potato. Gard. Chr. 70: 274–75.

BRIGGLE, L. W., 1953. A comparison of cytoplasmic-genotypic interactions in a group of cytoplasmic male sterile corn types. Ph. D. Thesis, Iowa State College, Ames. (unpublished).

BRIGGLE, L. W., 1956. Interaction of cytoplasm and genes in male sterile corn crosses involving two inbred lines. Agron. J. 48: 569–73.

BRIGGLE, L. W., 1957. Interactions of cytoplasm and genes in a group of male sterile corn types. Agron. J. 49: 543–47.

BROCK, R. D., 1954. Fertility in *Lilium* hybrids. Heredity 8: 409–18.

BURTON, G. W., 1943. Interspecific hybrids in the genus *Paspalum* (Dallis grass). J. Hered. 34: 15–23.

BURTON, G. W., 1948. The method of reproduction in common Bahia grass, *Paspalum notatum*. J. Amer. Soc. Agron. 40: 443–52.

BURTON, G. W., 1958. Cytoplasmic male sterility in pearl millet (*Pennisetum glaucum*). Agron. Jour. 50: 230.

CAMERON, D. R. and R. MOAV, 1957. Inheritance in *Nicotiana tabacum*. XXVII. Pollenkiller, an alien genetic locus inducing abortion of microspores not carrying it. Genetics 42: 326–35.

CHILDERS, W. R., 1952. Male sterility in *Medicago sativa*. Sci. Agric. 32: 351–64.

CHITTENDEN, F. J., 1914. The rogue wallflower. J. Bot. 52: 265–71.

154   S. K. JAIN

CHITTENDEN, F. J., 1927. Cytoplasmic inheritance in flax. J. Hered. 18: 337–43.
CHITTENDEN, F. J. and C. PELLEW, 1927. A suggested interpretation of certain cases of Anisogeny. Nature 119: 10–11.
CHRISTOFF, M., 1932. Male sterility in *Nicotiana*. Proc. VI Int. Cong. Genet. Ithaca: 20.
CLAASSEN, C. E. and A. HOFFMAN, 1950. The inheritance of pistillate character in castors and its possible utilization in the production of hybrid seed. Agron. J. 42: 79–82.
CLARKE, A. E. and J. R. FRYER, 1930. Seed setting in alfalfa. Sci. Agric. II: 138–43.
CLARKE, A. E. and L. H. POLLARD, 1949. The amount of selfpollination in male sterile onion lines. Proc. Amer. Soc. Hort. Sci. 53: 299–301.
CLARKE, F. J., 1942. Cytological and genetic studies of sterility in inbred and hybrid maize. Conn. Agric. Exp. Sta. Bull. 465: 705–26.
CLAUSEN, J., 1926. Genetical and cytological investigations on *Viola tricolor* L. and *V. arvensis* Murr. Hereditas 8: 1–156.
CLAUSEN, J., 1930. Male sterility in *Viola orphanides*. Hereditas 14: 53–72.
CLAYTON, E. E., 1950. Male sterile tobacco. J. Hered. 41: 171–75.
CONNORS, C. H., 1926. Sterility in peaches. Mem. Hort. Soc. New York 3: 215–22.
COOPER, D. C., 1952. The transfer of DNA from the tapetum to the microsporocytes at the onset of meiosis. Amer. Natur. 86: 219–29.
CORRENS, C., 1908. Die Rolle der männlichen Keimzellen bei der Geschlechtsbestimmung der gyndioecischen Pflanzen. Ber. Deut. Bot. Ges. 36: 686–701.
CORRENS, C., 1926. Uber Fragen der Geschlechtsbestimmung bei höheren Pflanzen. Zschr. ind. Abst. Vererb. 41: 5.
CORRENS, C., 1928. Bestimmung, Vererbung und Verteilung des Geschlechts bei den höheren Pflanzen. Handb. Vererb. wiss. 2: Pp. 138.
CRANE, M. B., 1915. Heredity of types of inflorescence and fruits in the tomato. J. Genetics 5: 1–11.
CRANE, M. B. and W. J. C. LAWRENCE, 1931. Inheritance of sex, color and hairiness in *Rubus idaeus* L. J. Genetics 24: 243–55.
CRANE, M. B. and P. T. THOMAS, 1949. Reproductive versatility in *Rubus*. III. Raspberry-blackberry hybrids. Heredity 3: 99–107.
CROWDER, L. V., 1953. Interspecific and intergeneric hybrids of *Festuca* and *Lolium*. J. Hered. 44: 195–203.
CURRENCE, T. M., 1933. Nodal sequence of flower type in cucumber. Proc. Amer. Soc. Hort. Sci. 29: 477–79.
CURRENCE, T. M., 1944. A combination of semisterility with two simply inherited characters that can be used to reduce the cost of hybrid tomato seed. Proc. Amer. Soc. Hort. Sci. 44: 403–06.
DAHLGREN, K. V. O., 1932. Über eine Form von *Primula officinalis* mit pistilloiden Staufgefassen und ihre Vererbung. Hereditas 17: 115–30.
DARLINGTON, C. D., 1937. Recent Advances in Cytology. 2nd. ed.: Pp. 671.
DARLINGTON, C. D., 1946. Evolution of genetic systems. 2nd ed.: Pp. 149.
DARLINGTON, C. D. and A. HAQUE, 1955. The timing of mitosis and meiosis

in *Allium ascalonicum*. A problem of differentiation. Heredity 9: 117–27.

DARROW, G. M., 1927. Sterility and fertility in the strawberry. J. Agric. Res. 34: 393–411.

DARWIN, C., 1890. Variation of animals and plants under domestication. p. 149.

DAVIDSON, F. R., H. E. BREWBAKER and N. A. THOMPSON, 1924. Brittle straw and other abnormalities in rye. J. Agric. Res. 28: 169–72.

DELASTAING, N., 1954. (Contribution to the cytology of the Genus *Salvia*). Rev. Cytol. Biol. Veg. 15: 195–236.

DENISEN, E. L. and E. S. HABER, 1950. Maleic hydrazide on sweet corn. North Central Weed Control Conf. Res. Rep.: Pp. 147.

DETJEN, L. R., 1927. Sterility in the common cabbage (*Brassica oleracea*). Mem. Hort. Soc. New York 3: 277–80.

DODDS, K. S. and N. W. SIMMONDS, 1946a. Genetical and cytological studies of *Musa*. VIII. The formation of polyploid spores. J. Genetics 47: 223–41.

DODDS, K. S. and N. W. SIMMONDS, 1946b. A cytological basis of sterility in *Tripsacum laxum* N. Ann. Bot. N.S. 10: 109–16.

DORSEY, M. J., 1914. Pollen development in the grape with special reference to sterility. Minn. Agric. Exp. Sta. Bull. 144: Pp. 60.

DORSEY, M. J., 1915. Pollen sterility in grape. J. Hered. 6: 243.

DORSEY, M. J., 1919. A study of sterility in the plum. Genetics 4: 417–88.

DORST, J. C., 1952. A questionable novum. Euphytica 1: 81–83.

DUVICK, D. N., 1956. Allelism and comparative genetics of fertility restoration of cytoplasmically pollen sterile maize. Genetics 41: 544–65.

EAST, E. M., 1932. Studies on selfsterility. IX. The behaviour of crosses between selffertile and selfsterile plants. Genetics 17: 175–102.

EDWARDSON, J. R., 1955. The restoration of fertility to cytoplasmic male-sterile corn. Agron. J. 47: 457–461.

EDWARDSON, J. R., 1956. Cytoplasmic male sterility. Bot. Rev. 22: 696–738.

ELLISON, W., 1936. Meiosis and fertility in certain British varieties of the cultivated potato. Genetica 18: 217–54.

EMERSON, R. A., G. W. BEADLE and A. C. FRASER, 1935. A summary of linkage studies in maize. Cornell Univ. Agric. Exp. Sta. Mem. 180: Pp. 83.

EMERSON, S. H., 1938. The genetics of selfincompatibility in *Oenothera organensis*. Genetics 23: 190–202.

ERLANSON, E. W., 1931. Sterility in wild roses and in some species hybrids. Genetics 16: 75–96.

ESKEW, E. B. and C. J. WILLARD, 1950. Maleic hydrazide on corn. North Central Weed Control Conf. Res. Rep.: Pp. 187.

EYSTER, L. A., 1921. Heritable characters of maize. VII. Male-sterile. J. Hered. 12: 138–41.

FABERGÉ, A. C., 1937. The cytology of the male sterile *Lathyrus odoratus*. Genetica 19: 423–30.

FORD, L. E., 1950. A genetic study involving male sterility in *Coleus*. Genetics 35: 664.

FOSKETT, R. L., 1954. Nature and inheritance of male sterility in the onion variety Scott Country Globe. Iowa State Coll. J. Sci. 28: 317.

156   S. K. JAIN

FRANKEL, O. H., 1940. Studies in *Hebe*. II The significance of male sterility in the genetic system. J. Genetics 40: 171–84.

FRANKEL, R., 1956. Graft-induced transmission to progeny of cytoplasmic male sterility in *Petunia*. Science 124: 684–85.

FUKASAWA, H., 1953. Studies on restoration and substitution of nucleus in *Aegilotricum*. I. Appearance of male-sterile *durum* in substitution crosses. Cytologia 18: 167–75.

FUKASAWA, H., 1954. On the free aminoacids in anthers of male-sterile wheat and maize. Jap. J. Genet. 29: 135–37.

FUKASAWA, H., K. MITO and M. FUJIWARA, 1957. Preventative effect of sugars on the pollen degeneration of wheat plant. Bot. Mag. Tokyo 70: 251–57.

FUKUDA, Y., 1927. Cytological studies on the development of pollen grains in different races of *Solanum tuberosum* with special reference to sterility. Bot. Mag. Tokyo 61: 459–76.

GABELMAN, W. H., 1949. Reproduction and distribution of the cytoplasmic factor for male sterility in maize. Proc. Nat. Acad. Sci. (Wash.) 35: 635–40.

GABELMAN, W. H., 1950. Characteristics of the cytoplasmic factor for male sterility in maize. Genetics 35: 665.

GAGNIEU, A., 1951. (Pollen production in the apple: Possibility of monofactorial genic lethality). Ann. Inst. Nat. Rech. Agron. 1: 455–96.

GAIRDNER, A. E., 1929. Male sterility in flax. II. A case of reciprocal crosses differing in F_2. J. Genetics 21: 117–24.

GAJEWSKI, W. A., 1937. A contribution to the knowledge of the cytoplasmic influence on the effect of nuclear factors in *Linum*. Acta Soc. Bot. Polon. 14: 205–14.

GARBER, R. J., 1922. Inheritance and yield with particular reference to rust resistance and panicle types in oats. Minn. Agric. Exp. Sta. Techn. Bull. 7: 5–62.

GÄRTNER, C. F., 1844. Versuche der Beobachtungen über Befruchtungsorgane. Stuttgart: Pp. 97.

GATES, R. R., 1911. Pollen formation in *Oenonthera gigas*. Ann. Bot. 25: 909–40.

GREGORY, F. G. and O. N. PURVIS, 1947. Abnormal flower development in barley involving sex reversal. Nature Lond. 160: 221–22.

GREGORY, R. P., 1905. The abortive development of the pollen in certain sweat-peas. Proc. Camb. Phil. Soc. 13: 148–57.

GRELAUCH, V. A., 1951. The effect of maleic hydrazide on some water relations of plants. Literature Summary by ZUKEL (1954). Naugatuck Co., Rubber Division.

GUERIN, P., 1926. Le development de l'anthere chez les Gentianacees. Bull. Soc. Bot. de France 73: 5–18.

HALL, W. C., 1949, Effects of photoperiod and nitrogen supply on growth and reproduction in the gherkin. Pl. Physiol. 24: 753–69.

HARTE, C. and B. BISSINGER, 1952. Entwicklungsgeschichtliche Untersuchung der durch die Faktoren fr und ster bedingten Pollensterilität bei *Oenothera*. Zschr. ind. Abst. Vererb. 84: 251–69.

HAYES, H. K., F. R. IMMER and D. C. SMITH, 1955. Methods of plant breeding. 2nd ed.: 551 pp.

HEILBORN, O., 1932. Lethal gene combinations and pollen sterility in diploid apple varieties. A critique theory. Hereditas 16: 1–18.

HERIBERT-NILSSON, N., 1913. Potatisfordaling och potatisbedomsang. W. Weibull. Årsbok 8: 4–31.

HESLOP-HARRISON, J., 1957. The experimental modification of sex expression in flowering plants. Biol. Revs. 32: 38–90.

HESLOP-HARRISON, J. and Y. HESLOP-HARRISON, 1958. Longday and auxin induced male sterility in Silene pendula L. Port. Acta Biol. Ser. A, 5: 79–94.

HIORTH, G., 1948. Über Hemmungssysteme bei Godetia Whitneyi. I. Zschr. ind. Abst. Vererb. 82: 12–63.

HOGABOAM, G. J., 1957. Factors influencing phenotypic expression of cytoplasmic male sterility in the sugarbeet (Beta vulgaris L.) J. Amer. Soc. Sugarbeet Technol. 9: 457–65.

HOLDEN, J. W. and M. MOTA, 1956. Nonsynchronised meiosis in binucleate pollen mother cells of an Avena hybrid. Heredity 10: 109–17.

HOWLETT, F. S., 1936. The effect of carbohydrate and of nitrogen deficiency upon microsporogenesis and the development of the male gametophyte in the tomato. Ann. Bot. 50: 767–804.

HSU, K., 1945. On sterility resulting from crossing different types of rice. Ind. J. Genet. Pl. Breed. 5: 51–57.

ISENBERG, F. M. R., M. L. ODLAND, H. W. POPP and C. O. JENSEN, 1951. The effect of maleic hydrazide on certain dehydrogenases in tissues of onion plants. Science 113: 58–60.

ISHIKAWA, J., 1927. (Studies on the inheritance of sterility in rice). J. Coll. Agr. Hokkaido Imp. Univ. 20: 79–201.

JAIN, S. K., 1956. Natural incidence of male sterility and its chemical induction in crop plants. Assoc. Diss., I.A.R.I., N. Delhi.

JANAKI-AMMAL, E. K., 1941. The breakdown of meiosis in a male sterile Saccharum. Ann. Bot. N.S. 5: 83–87.

JANAKI-AMMAL, E. K., 1942. Intergeneric hybrids of Saccharum. IV. Saccharum-Narenga. J. Genetics 44: 23–32.

JARETZKY, R., 1927. Die Degenerationserscheinungen in der Blüten von Rumex flexuosus. Forst. Jahrb. Wiss. Bot. 66: 300–20.

JASMIN, J. J., 1954. Male sterility in Solanum melongena L. Preliminary report. Proc. Amer. Soc. Hort. Sci. 63: 443.

JENKIN, T. J., 1930. Selffertility in perennial ryegrass (Lolium perenne L.). Welsh. Pl. Breed. Sta. 12: 100–19.

JOHNSSON, H., 1944. Meiotic aberrations and sterility in Alopecurus myusroides. Hereditas 30: 469–566.

JONES, D. F., 1934. Unisexual maize plants and their bearing on sex differentiation in other plants and animals. Genetics 19: 552–67.

JONES, D. F., 1950. The interrelation of plasmagenes and chromogenes in pollen production in maize. Genetics 35: 507–12.

JONES, D. F., 1951. The cytoplasmic separation of the species. Proc. Nat. Acad. Sci. (Wash.) 37: 408–10.

JONES, D. F. and P. C. MANGELSDORF, 1951. The production of hybrid seedcorn without detasseling. Conn. Agric. Exp. Sta. Bull. 550.

JONES, D. F., H. T. STINSON and U. KHOO, 1957. Transmissible variations in the cytoplasm within species of higher plants. Proc. Nat. Acad. Sci. (Wash.) 43: 598–602.

JONES, H. A. and A. E. CLARKE, 1943. Inheritance of male sterility in the onion and the production of hybrid seed. Proc. Amer. Soc. Hort. Sci. 43: 189–94.

JONES, H. A. and G. N. DAVIS, 1944. Inbreeding and heterosis and the relations to the development of new varieties of onion. U.S.D. A. Tech. Bull. 874: Pp. 28.

JONES, H. A. and S. L. EMSWELLER, 1937. A male sterile onion. Proc. Amer. Soc. Hort. Sci. 34: 583–85.

JOSEPHSON, L. M. and M. T. JENKIN, 1948. Male sterility in corn hybrids. J. Amer. Soc. Agron. 40: 267–74.

JUNGENHEIMER, R. W., 1951. Evaluation of genetic male sterility in the hybrid corn program. Rep. VI Hybrid Corn Ind. Res. Conf.: 25–28.

KAJJARI, N. B. and V. M. CHAVAN, 1953. A male-sterile jowar. Ind. J. Genet. Pl. Breed. 13: 48–49.

KAPPERT, H., 1944. Untersuchungen uber Plasmonwirkungen bei *Aquilegia*. Gynodioecie-Heterosis-Gestalt des Sporus. Flora 37: 95–105.

KARPER, R. E. and J. C. STEPHENS, 1936. Floral abnormalities in *Sorghum*. J. Hered. 27: 183–94.

KIHARA, H., 1951. Substitution of nucleus and its effects on genome manifestation. Cytologia 16: 177–93.

KIHARA, H. and N. KONDO, 1943. (Studies in amphidiploids of *Aegilops caudata* × *Ae. umbellulata* induced by colchicine). Seiken Ziho 2: 24–42.

KLEY, F. K. VAN DER, 1954. Male sterility and its importance in breeding heterosis varieties. Euphytica 3: 117–24.

KLEY, F. K. VAN DER, 1955. The occurrence and rates of reproduction of various male sterility genes. Genetica 27: 453–64.

KOOPMANS, A., 1951. Cytogenetic studies on *Solanum tuberosum* L. and some of its relatives. Genetica 25: 193–337.

KOOPMANS, A., 1952. Changes in sex on the flowers of the hybrid *Solanum rybinii* × *S. chacoense*. Genetica 26: 359–80.

KOOPMANS, A., 1954. Changes in sex in the flowers of the hybrid *Solanum rybinii* × *S. chacoense* II. Plastmatic influences upon gene action. Genetica 27: 273–85.

KOOPMANS, A., 1955. Changes in sex in the flowers of the hybrid *Solanum rybinii* × *S. chacoense*. III. Data about the reciprocal cross *Solanum chacoense* × *S. rybinii*. Genetica 27: 465–471.

KOOPMANS, A., 1959. Changes in sex in the flowers of the hybrid *Solanum rybinii* × *S. chacoense*. IV. Further data on the reciprocal hybrid *S. rybinii* × *S. chacoense*. Genetica 30: in press.

KOSTOFF, D., 1930. Chromosomal aberrants and gene mutations in *Nicotiana* by grafting. J. Genet. 22: 399–418.

KNOWLTON, H. E., 1924. Pollen abortion in the peach. Proc. Amer. Soc. Hort. Sci. 21: 67.

KRAUS, E. J. and M. R. KRAYBILL, 1918. Vegetation and reproduction with special reference to the tomato. Oregon Agric. Exp. Sta. Bull. no. 149.

KRISHNASWAMY, N. and G. N. R. AYYANGAR, 1942. Certain abnormalities in millets induced by X-rays. Proc. Ind. Acad. Sci. Sect. B. 16: 1–9.

KÜHN, E., 1937. Zur Physiologie der Pollenkeimung bei *Matthiola*. Planta 27: 304–33.

KVAALE, E., 1927. Abortive and sterile apple pollen. Mem. Hort. Soc. New York 3: 399–408.

KYONO, H., 1955. (Abnormalities in pollen grain formation in *Trillium kamtschaticum* Pall.). La Kromosomo 22–24: 826–29.

LACOUR, L. F., 1949. Nuclear differentiation in the pollen grain. Heredity 3: 319–37.

LAIBACH, F. and F. J. KRIBBEN, 1950. Der Einfluss von Wuchsstoff auf die Blutenbildung der Gurke. Naturwissenschaften 37: 114.

LAMM, R., 1945. Cytogenetic studies on *Solanum* sect. *Tuberarium*. Hereditas 31: 1–128.

LAMM, R., 1953. Investigations on some tuberbearing *Solanum* hybrids. Hereditas 39: 97–112.

LARSON, R. E. and S. PAUR, 1948. The description and inheritance of a functionally sterile flower mutant in the tomato and its possible value in hybrid tomato seed production. Proc. Amer. Soc. Hort. Sci. 52: 355–64.

LEIGHTY, C. E. and W. J. SANDO, 1924. Pistillody in wheat flowers. J. Hered. 15: 253–68.

LEOPOLD, A. C. and W. H. KLEIN, 1951. Maleic hydrazide as an antiauxin in plants. Science 114– 9–10.

LESLEY, J. W. and M. M. LESLEY, 1939. Unfruitfulness in the tomato caused by male sterility. J. Agric. Res. 58: 621–30.

LESLEY, J. W. and M. M. LESLEY, 1955. The effect in T, R_1 and R_2 of treating tomato seeds with X-ray and P^{32}. Genetics 40: 581.

LEVAN, A., 1935. Zytologische Studien an *Allium schoenoprasum*. Hereditas 22: 1–126.

LEVAN, A., 1941. Syncyte formation in the pollen mother cells of haploid *Phleum pratense*. Hereditas 27: 243–62.

LEWIS, D., 1939. Genetical studies in cultivated raspberries. I. Inheritance and linkage. J. Genetics 38: 367–79.

LEWIS, D., 1941. Male sterility in natural populations of hermaphrodite plants. New Phytologist 40: 56–63.

LEWIS, D., 1942. The evolution of sex in flowering plants. Biol. Revs. 17: 46–67.

LEWIS, D. and L. K. CROWE, 1952. Male sterility as an outbreeding mechanism in *Origanum vulgare*. Heredity 6: 136.

LEWIS, D. and L. K. CROWE, 1956. The genetics and evolution of gynodioecy. Evolution 10: 115–25.

LIMA-DE-FARIA, A., 1947. Disturbances in microspore cytology of *Anthoxanthum*. Hereditas 33: 539–51.

LINDSTRÖM, E. W., 1933. Hereditary radium-induced variations in the tomato. J. Hered. 24: 129–37.

LITTLE, T. M., H. A. JONES and A. E. CLARKE, 1944. The distribution of the male sterility gene in varieties of onion. Herbertia II: 310–12.

LIVINGSTON, C., M. G. PAYNE and J. L. FULTS, 1954. Effects of maleic hydrazide and 2–4, dichlorophenoxyacetic acid on the free aminoacids in sugarbeets. Bot. Gaz. 116: 148–56.

LODEN, H. D. and T. R. RICHMOND, 1951. Hybrid vigour in cotton. Cytogenetic aspects and practical applications. Econ. Bot. 5: 387–408.

LOEHWING, W. F., 1938. Physiological aspects of sex in Angiosperms. Bot. Rev. 4: 581–625.

LONGLEY, A. E., 1927. Relationship of polyploidy to pollen sterility in the genera *Rubus* and *Fragaria*. Mem. Hort. Soc. New York 3: 15–17.

LONGLEY, A. E. and C. F. CLARK, 1930. Chromosome behaviour and pollen production in the potato. J. Agric. Res. 41: 867–88.

LOVELESS, A., 1952. Chemical and biochemical problems arising from the study of chromosome breakage by alkylating agents and heterocyclic compounds. Heredity 6 Suppl.: 293–98.

LOWIG, E., 1928. (Sterility in some "good" species-*Secale montanum* and *Iris* spp.). Flora 23: 62–104.

MARSDEN-JONES, E. M. and W. B. TURRILL, 1925. Studies in *Ranunculus*. III. Further experiments concerning sex in *R. acris*. J. Genetics 31: 363–78.

MARTIN, J. A. and J. H. CRAWFORD, 1951. Several types of sterility in *Capsicum frutescens*. Proc. Amer. Soc. Hort. Sci. 57: 335–39.

MATHER, K., 1940. Outbreeding and separation of the sexes. Nature 145: 484.

MICHAELIS, P., 1933. Entwicklungsgeschichtlich-genetische Untersuchung an *Epilobium*. II. Die Bedeutung des Plasma für die Pollenfertilität des *E. luteum-hirsutum* Bastards. Zschr. ind. Abst. Vererb. 65: 1–71.

MICHAELIS, P., 1954. Cytoplasmic inheritance in *Epilobium* and its theoretical significance. Adv. Genet. 6: 288–402.

MICHAELIS, P. and G. MICHAELIS, Über die Konstanz des Zytoplasmons bei *Epilobium*. Planta 35: 467–512.

MILLER, J. C., 1929. A study of some factors affecting seedstalk development in cabbage. Cornell Agric. Exp. Sta. Bull. 488.

MOFFETT, A. A., 1932. Studies on the formation of multinuclear giant pollen grains in *Kniphofia*. J. Genetics 25: 315–37.

MOH, C. C. and R. A. NILAN, 1953. Multiovary in barley. A mutant induced by atomic bomb irradiation. J. Hered. 44: 183–84.

MOL, W. E. DE, 1933. Die Entstehungsweize anormaler Pollenkörner bei Hyazinthen, Tulipen und Narzissen. Cytologia 5: 31–65.

MONOSMITH, H. R., 1926. Male sterility in *Allium cep*. Doctoral Thesis, Univ. of Calif. (unpublished).

Moore, R. H., 1950. Several effects of maleic hydrazide on plants. Science 112: 52–53.

Morris, R., 1952. Transmitted pollen and chromosomal aberrancies induced in maize by tassel exposure in a nuclear reactor. Amer. J. Bot. 39: 452–57.

Müntzing, A., 1932. Apomictic and sexual seed formation in *Poa*. Hereditas 17: 131–54.

Müntzing, A., 1946. Sterility in rye populations. Hereditas 32: 521–49.

Myers, W. M., 1946. Effects of cytoplasm and gene dosage on expression of male sterility in *Dactylis glomerata*. Genetics 31: 225–26.

McFarlane, E. W. E., 1950. Somatic mutations produced by organic mercurials in flowering plants. Genetics 35: 122–23.

McIlrath, W. J., 1950. Response of the cotton plant to maleic hydrazide. Amer. J. Bot. 37: 816–19.

McIlrath, W. J., 1953. Maleic hydrazide induced male sterility in *Sorghum*. Liter. Summary by Zukel, Naugatuck Co., Rubber Division.

Nagai, F., 1926. Studies on the mutations in *Oryza sativa* L. Jap. J. Bot. 3: 55–56.

Nakamura, M., 1936. Experimental and cytological studies on the instability of the meiotic division of the pollenmothercells of *Impatiens Balsamina* L. caused by the effect of high air temperature. Mem. Fac. Sci. Agr. Taikohu Imp. Univ. 17: Pp. 121.

Nakamura, M., 1943. Cytological and ecological studies of the genus *Citrus*, with special reference to the occurrence of the sterile grains. Mem. Fac. Sci. Agr. Taihoku Imp. Univ. 37: 53–159.

Naylor, A. W., 1950. Observations on the effects of maleic hydrazide on flowering of tobacco, maize and cocklebur. Proc., Nat. Acad. Sci. (Wash.) 36: 230–32.

Naylor, F. L., 1941. Effect of length of induction period on floral development in *Xanthium pennysylvanicum*. Bot. Gaz. 103: 146–000.

Nicolas, J. H., 1927. Sterility encountered in rose breeding. Mem. Hort. Soc. New York 3: 55–57.

Nielsen, E. L., 1955. Cytological disturbances influencing fertility in *Bromus inermis*. Bot. Gaz. 116: 293–305.

Noack, K. L., 1932. Über *Hypericum*-kreuzungen. II. Beobachtungen an *Hypericum* Artbastarden. Ber. Deut. Bot. Ges. 50: 256–68.

Nördenskiöld, H., 1945. Cytogenetic studies in the genus *Phleum*. Acta Agric. Sv., Stockholm. 1: 1–137.

Oehlkers, F., 1927. Entwicklungsgeschichte der pollensterilität einigen Oenotheren. Zschr. ind. Abst. Vererb. 43: 265–84.

Oehlkers, F., 1938. Bastardierungsversuche in der Gattung *Streptocarpus* Lindl. I. Plasmatische Vererbung und die Geschlechtsbestimmung von Zwitterpflanzen. Zschr. Bot. 32: 305–93.

Oehlkers, F., 1952. Neue Uberlegungen zum Problem der auszerkaryotischen Vererbung. Zschr. ind. Abst. Vererb. 84: 213:250.

Offerijns, F. J. M., 1938. Meiosis in the pollen mother cells of *Canna glauca"*, Pure Yellow". Genetica 20: 59–65.

OLDEMEYER, R. K., 1957. Sugarbeet male sterility. J. Amer. Soc. Sugarbeet Techn. 9: 381–86.

OLMO, H. P., 1943. Pollination of the Almeria grape. Proc. Amer. Soc. Hort. Sci. 42: 401–06.

OSTENFELD, C. H., 1906. Castration and hybridization experiments with some species of *Hieracium*. Bot. Tidskr. 27: 225–48.

OWEN, F. V., 1945. Cytoplasmically inherited male sterility in sugarbeets. J. Agric. Res. 71: 423–40.

OWEN, F. V., 1948. Utilization of male sterility in breeding superior yielding sugarbeets. Proc. Amer. Soc. Sugarbeet Tech. 5: 156–61.

OWEN, F. V., 1952. Mendelian male sterility in sugarbeets. Proc. Amer. Soc. Sugarbeet Tech. 7: 371–76.

PAINTER, T. S., 1943. Cell growth and nucleic acids in the pollen of *Rhoeo discolor*. Bot. Gaz. 105: 58–68.

PARKEY, W., 1957. Cytoplasmic influence on the production of the pistillate sex expression in castor beans. Agron. J. 49: 427–28.

PATE, J. B. and J. F. JOYNER, 1958. The inheritance of a male sterility factor in kenaf (*Hibiscus cannabinus L.*) Agron. Jour. 50: 402–03.

PERSSON, A. and L. RAPPAPORT, 1958. Gibberellin-induced systemic fruit set in a male sterile tomato. Science 127: 816–17.

PETERSON, C. E. and R. L. FOSKETT, 1953. Occurrence of pollen sterility in seed fields of Scott Country Globe onions. Proc. Amer. Soc. Hort. Sci. 62: 443–48.

PHIPPS, I. F., 1928. Heritable characters in maize. XXXI. Tassel seed–4. J. Hered. 19: 399–404.

POWERS, L. and E. J. GARDNER, 1945. Frequency of aborted pollen grains and microcytes in guyale (*Parthenium argentatum* Gray.). J. Amer. Soc. Agron. 37: 184–93.

PUNNETT, R. C., 1932. Further studies of linkage in the sweetpea. J. Genetics 26: 97–112.

PUTT, E. D., 1954. Cytogenetic studies of sterility in rye. Canad. J. Agri. Sci. 34: 81–119.

RAEBER, J. G. and A. BOLTON, 1955. A new form of male sterility in *Nicotiana tabacum* L. Nature, Lond. 176: 314–15.

RAMAER, H., 1935. Cytology of *Hevea*. Genetica 17: 193–236.

RAMAN, V. S., 1955. Cytology of Indian Jasmines. III. Meiosis. Cytologia 20: 133–47.

RAMANUJAM, S., 1935. Male sterility in rice. Madras Agric. J. 13: 190.

REHM, S., 1952. Male sterile plants by chemical treatment. Nature 170: 38–39.

RENNER, O., 1919. Zur Biologie und Morphologie der männlichen Haplonten einiger Oenotheren. Zschr. Bot. II: 306–80.

RHEDER, A., 1911. Pistillody in stamens of *Hypericum nudiflorum*. Bot. Gaz. 51: 230–31.

RHOADES, M. M., 1931. Cytoplasmic inheritance of male sterility in *Zea mays*. Science 73: 340–41.

RHOADES, M. M., 1933. The cytoplasmic inheritance of male sterility in *Zea mays*. J. Genetics 27: 71–93.

RHOADES, M. M., 1943. Genic induction of an inherited cytoplasmic difference. Proc. Nat. Acad. Sci. (Wash.) 29: 327–29.

RHOADES, M. M., 1950. Gene induced mutation of a heritable cytoplasmic factor producing male sterility in maize. Proc. Nat. Acad. Sci. (Wash.). 36: 634–35.

RICHEY, F. D. and G. F. SPRAGUE, 1932. Some factors affecting the reversal of sex expression in the tassels of maize. Amer. Natur. 66: 433–43.

RICK, C. M., 1944. A new male-sterile mutant in the tomato. Science 99: p. 543.

RICK, C. M., 1945. A survey of cytogenetic causes of unfruitfulness in the tomato. Genetics 30: 347–62.

RICK, C. M., 1947. The effect of planting design upon the amount of seed produced by malesterile tomato plants as a result of natural crosspollination. Proc. Amer. Soc. Hort. Sci. 50: 274–84.

RICK, C. M., 1948. Genetics and development of nine malesterile tomato mutants. Hilgardia 18: 599–633.

RICK, C. M., 1949. Rates of natural crosspollination of tomatoes in various localities in California as measured by the fruits and seeds on male-sterile plants. Proc. Amer. Soc. Hort. Sci. 54: 237–52.

RICK, C. M., 1956. Rept. Tomato Genetics Coop. 6: 26.

RIDDLE, O. C. and C. A. SUNESON, 1944. Crossing studies with male sterile barley. J. Amer. Soc. Agron. 36: 62–65.

RIFE, D. C., 1948. Simply inherited variations in *Coleus*. J. Hered. 39: 85–91.

ROEVER, W. E., 1948. A promising type of male sterility for use in hybrid tomato seed production. Science 107: 506.

ROGERS, J. S. and J. R. EDWARDSON, 1952. The utilization of cytoplasmic male sterile inbreds in the production of corn hybrids. Agron. J. 44: 8–13.

ROSA, J. T., 1928. Flower types in *Cucumis* and *Citrullus*. Hilgardia 3: 235–50.

ROSCOE, M. V., 1927a. Cytological studies in the genus *Wisteria*. Bot. Gaz. 84: 171–86.

ROSCOE, M. V., 1927b. Meiotic irregularities in a gigas form of *Potentilla anserina*. Bot. Gaz. 84: 307–16.

SALAMAN, R. N., 1910. Male sterility in potatoes a dominant Mendelian character, J. Linn. Soc. 39: 301–12.

SALAMAN, R. N., 1912. The hereditary characters in the potato. J. Roy. Hort. Soc. 38: p. 35.

SALAMAN, R. N. and J. W. LESLEY, 1922. Genetic studies in potatoes: Sterility. J. Agric. Sci. 12: 31–39.

SANSOME, F. W., 1936. Some experiments with *Geranium* species. J. Genetics 33: 359–63.

SAUNDERS, A. P. and G. L. STEBBINS, 1936. Cytogenetic studies in *Paeonia*. I. The compatibility of the species and the appearance of the hybrids. Genetics 23: 65–82.

SCHAFFNER, J. H., 1925. Experiments with various plants to produce change of sex in the individual. Bull. Torr. Bot. Club 52: 35–47.

SCHAFFNER, J. H., 1927. Sex and sex determination in the light of observations and experiments on dioecious plants. Amer. Natur. 61: 319–32.

SCHAFFNER, J. H., 1929. Progeny resulting from selfpollination of staminate plant of *Morus alba* showing sex reversal. Bot. Gaz. 87: 653.

SCHNELL, L. O., 1948. A study of meiosis in the microsporocytes of interspecific hybrids of *Solanum demissum* × *S. tuberosum* carried through from back-crosses. J. Agric. Res. 76: 185–212.

SCHOENE, D. L. and O. L. HOFFMANN, 1949. Maleic hydrazide. A unique growth regulant. Science 109: 90.

SCHÜRHOFF, P. N., 1922. Die Teilung des vegetativen Pollenkerns bei *Eichhornia crassipes*. Ber. Deut. Bot. Ges. 40: 60–63.

SCHWARTZ, D., 1951. The interaction of nuclear and cytoplasmic factors in the inheritance of male sterility in maize. Genetics 36: 676–96.

SCHWEMMLE, J., 1928. Genetischen und cytologischen Untersuchungen bei *Oenothera*. Jahrb. Wiss. Bot. 67: 849–76.

SCOTT, D. H. and M. E. RINER, 1946. Inheritance of male sterility in winter squash. Proc. Amer. Soc. Hort. Sci. 47: 375–77.

SCOTT, D. H. and J. H. WEINBERGER, 1944. Inheritance of pollen sterility in some peach varieties. Proc. Amer. Soc. Hort. Sci. 45: 229–32.

SCOTT, G. W., 1933. Sex ratios and fruit production studies in bush pumpkins. Proc. Amer. Soc. Hort. Sci. 30: 520–25.

SELIM, A. G., 1931. A cytological study of the *Oryza sativa* L. Cytologia 2: 2: 1–26.

SHIFRISS, O., 1945. Male sterilities and albino seedling in cucurbits. J. Hered. 36: 47–52.

SHIFRISS, O., 1956. Sex instability in *Ricinus*. Genetics 41: 265–80.

SHULL, G. H., 1923. Further evidence of linkage with crossing over in *Oenothera*. Genetics 8: 154–67.

SINGLETON, W. R. and D. F. JONES, 1930. Heritable characters of maize. XXXV. Male sterile. J. Hered. 21: 266–68.

SIRKS, M. J., 1924. Die gynanthere Form des Goldlacks und ihre Vererbung. Genetica 6: 537–48.

SIRKS, M. J., 1938. Plasmatic inheritance. Bot. Rev. 4: 113–31.

SKOOG, F., 1953. Substances involved in normal growth and differentiation. VI Symp. Abnormal and Pathological Plant Growth. Brookhaven Nat. Lab.: 1–21.

SMITH, C. O., 1927. A pistillate *Prunus*. J. Hered. 18: 537–41.

SMITH, L., 1938. Cytogenetic studies in *Triticum monococcum*. Genetics 23: p. 168

SMITH, L., 1939. Mutants and linkage studies in *Triticum monococcum* and *T. aegilopoides*. Mo. Agric. Exp. Sta. Res. Bull. 298.

SNOAD, B., 1954. Abortive meiosis in plasmodial PMC's of *Helianthemum*. Ann. Bot. 18: 1–6.

SOROKIN, H., 1927. Cytological and morphological investigations on gynodimorphic and normal forms of *Ranunculus acris* L. Genetics 12: 59–83.

SPRAGUE, G. F., 1939. Heritable characters in maize. L. Vestigial glume. J. Hered. 30: 143–45.

STEBBINS, G. L., 1958. The inviability, weakness and sterility of interspecific hybrids. Adv. Gen. 9: 147–216.

STEPHENS, J. C., 1937. Male sterility in *Sorghum*. Its possible utilization in production of hybrid seed. J. Amer. Soc. Agron. 29: 690–96.

STEPHENS, J. C. and R. F. HOLLAND, 1954. Cytoplasmic male sterility for hybrid sorghum seed production. Agron. J. 46: 20–23.

STEPHENS, J. C., J. H. KUYKENDALL and D. W. GEORGE, 1952. Experimental production of hybrid sorghum seed with a three way cross. Agron. J. 44: 367–73.

STEPHENS, J. C. and J. R. QUINBY, 1945. The Ms AV linkage group in sorghum. J. Agric. Res. 70: 209–18.

STEPHENS, S. G., 1942. Colchicine induced polyploids in *Gossypium*. J. Genetics 44: 272–95.

STOREY, W. B., 1953. Genetics of the papaya. J. Hered. 44: 70–78.

STOUT, A. B., 1923. Alternation of sexes and intermittent production of fruit in the spider flower. Amer. J. Bot. 10: 57–66.

STOUT, A. B. and C. F. CLARK, 1924. Sterilities of wild and cultivated potatoes with reference to breeding from seed. U.S.D.A. Bull. 1195: Pp. 32.

STOW, I., 1927. A cytological study on pollen sterility in *Solanum tuberosum* L. Jap. J. Bot. 3: 217–38.

STOW, I., 1930. Experimental studies on the formation of the embryosac-like giant pollen grains in the anthers of *Hyacinthus orientalis*. Cytologia 1: 417–39.

STRASBURGER, E., 1905. Die Apogamie der Eualchimillen und allgemeine Gesichtspunkte, die sich aus ihr ergeben. Jahrb. Wiss. Bot. 41: 88–164.

STRINGFIELD, G. H., 1958. Fertility restoration and yields in maize. Agron. Jour. 50: 215–18.

SUNESON, C. A., 1940. A male sterile character in barley: A new tool for the plant breeder. J. Hered. 31: 213–14.

SUNESON, C. A., 1945. The use of male sterility in barley improvement. J. Amer. Soc. Agron. 37: 72–73.

SUNESON, C. A. and B. R. HOUSTON, 1942. Male sterile barley for study of floral infection. Phytopath. 32: 431–32.

SUSA, T., 1927. Sterility in certain grapes. Mem. Hort. Soc. New York 3: 223–28.

SWAMINATHAN, M. S., 1952. Cytogenetic studies on *Solanum* species and hybrids. Doctoral Thesis, Cambridge (unpublished).

TAPLEY, W. T., 1923. The fruiting habit of the squash. Proc. Amer. Soc. Hort. Sci. 20: 312–19.

TAYLOR, J. H., 1950. The duration of differentiation in excised anthers. Amer. J. Bot. 37: 137–43.

TAYLOR, J. H., 1953. Autoradiographic detection of incorporation of P^{32} into chromosomes during meiosis and mitosis. Exp. Cell Res. 4: 164–73.

THOMAS, W. I. and I. J. JOHNSON, 1956. Inheritance of pollen restoration and transmission of cytoplasmic sterility in popcorn. Agron. J. 48: 472–74.

THOMPSON, J. M., 1927. A study in advancing gigantism with staminal sterility with special reference to the Lacythideae. Univ. Liverpool Publ. Hart. Bot. Lab. 4: 1–44.

TOKUMASU, S., 1951. (Male sterility in Japanese radish). Sci. Bull. Fac. Agric. Kyushu Univ. 13: 83–89'

TOURNOIS, J., 1911. Anomalies florales du Houblon japonais et du Chanvre determinées par des semis hâtifs. C. R. Acad. Sci. Paris 155: 297.

UPCOTT, M., 1937. Timing unbalance at meiosis in pollen sterile *Lathyrus odoratus*. Cytologia, Fujii Jubilee Volume: 299–310.

VALLEAU, W. P., 1918. Sterility in the strawberry. J. Agric. Res. 12: 613–70.

VILLERTS, A., 1942. Über die Verschiedenheit reziproker Art-Bastarde in der Gattung *Begonia*. J. Genetics 43: 223–36.

WARREN, F. S. and F. DIMMOCK, 1954. The use of chemicals and of male sterility to control pollen production in maize. Canad. J. Agric. Sci. 34: 48–52.

WEATHERWAX, P., 1925. Anomalies in maize and its relatives. III. Carpelloidy in maize. Bull. Torr. Bot. Club 52: 167–80.

WELCH, J. E. and E. L. GRIMBALL, 1947. Male sterility in the carrot. Science 109: p. 594.

WELZEL, G., 1954. Entwicklungsgeschichtlich-genetische Untersuchungen an pollensterilen Mutanten von *Petunia*. Zschr. ind. Abst. Vererb. 86: 35–53.

WESTERGAARD, M., 1958. The mechanism of sex determination in dioecious flowering plants. Adv. Gen. 9: 217–82.

WETTSTEIN, F. von, 1924. Über Fragen der Geschlechtsbestimmung bei Pflanzen Naturwissenschaften 12: 761–68.

WHITAKER, T. W., 1931. Sex ratio and sex expression in the cultivated cucurbits. Amer. J. Bot. 18: 359–66.

WHYTE, R. O., 1929. Studies in *Ranunculus*. II. The cytological basis of sex in *R. acris*. J. Genetics 21: 183–91.

WITTWER, S. H. and I. G. HILLYER, 1954. Chemical induction of male streility in cucurbits. Science 120: 893–94.

WOODWORTH, R. H., 1930. Cytological studies in the Betulaceae. III Parthenogenesis and polyembryony in *Alnus rugosa*. Bot. Gaz. 89: 402–09.

YOST, H. T., W. R. SINGLETON and A. F. BLAKESLEE, 1953. The effect of thermal neutron radiation on the chromosomes of *Datura*. Proc. Nat. Acad. Sci. (Wash) 39: 292–97.

YOUNG, W. J., 1923. The formation and degeneration of germ cells in the potato. Amer. J. Bot. 10: 325–35.